K.J. Bartlett
1995

BARTLETT
Marram House
2 St. Andrew's Drift
Langham, Holt
Norfolk NR25 7AG
Telephone: Binham (01328) 830696

HANDBOOK OF

Type
and
Lettering

DESIGN PRESS

HANDBOOK OF
Type and Lettering

Hildegard Korger

English translation by Ingrid Li
Typesetting by V & M Graphics, Inc.

Translation of *Schrift und Schrieben*, Sixth Edition
by Hildegard Korger
Copyright © 1986 Fachbuchverlag GmbH Leipzig
Translation copyright © 1992 by Design Press

First Edition, First Printing
Printed in the United States of America

Reproduction or republication of the content in any manner, without the express written permission of the publisher, is prohibited. The publisher takes no responsibility for the use of any of the materials or methods described in this book, or for the products thereof.

Library of Congress Cataloging-in-Publication Data

Korger, Hildegard.
　[Schrift und Schreiben. English]
　Handbook of type and lettering / Hildegard Korger.
　　　p.　　cm.
　Translation of: Schrift und Schreiben.
　Includes bibliographical references and index.
　ISBN 0-8306-1834-1
　1. Type and type-founding.　2. Lettering.
　I. Title
Z250.K63　　1991　　　　　　91-19575
686.2'24 — dc20　　　　　　　CIP

Design Press offers posters and The Cropper, a device for cropping artwork, for sale. For information, contact Mail-order Department. Design Press books are available at special discounts for bulk purchases for sales promotions, fund raisers, or premiums. For details contact Special Sales Manager. Questions regarding the content of the book should be addressed to:
　Design Press
　11 West 19th Street
　New York, NY 10011

Design Press books are published by Design Press, an imprint of TAB BOOKS. TAB BOOKS is a Division of McGraw-Hill, Inc. The Design Press logo is a trademark of TAB BOOKS.

First published in Great Britain in 1992 by
Lund Humphries Publishers Ltd
Park House
1 Russell Gardens
London NW11 9NN

British Library Cataloguing-in-Publication Data

A catalogue record for this book
is available from the British Library.

ISBN 0-85331-625-2

Dedicated to my teacher,
Professor Albert Kapr

Contents

Foreword *9*

1 Advice on Design

Introduction *13*
Basic Elements of Design *13*
  Contrast *13*
  Rhythm *13*
  Size Relationships *14*
  Color *16*
    Light/Dark Contrast *16*
    Warm/Cool Contrast *17*
    Pure and "Broken" Colors *17*
    Color Associations *17*
  Movement and Focal Points *18*
  Three-Dimensionality on a Flat Surface *19*
Optical Illusions in Lettering *20*
Letter, Word, and Line Spacing *23*
  Letter Spacing *23*
  Word Spacing *25*
  Line Spacing *25*
Composition *26*
  Principles of Composition *26*
  Layout *28*
Formats *29*
Punctuation Marks *30*

2 Introductory Course

Introduction *32*
The Work Space, Materials, and Tools *33*
  The Work Space *33*
  Paper *33*
  Ink *33*
  Pens and Nibs *34*
  Brushes *36*
  Other Writing Tools *37*
  Drawing Materials *37*
Exercises *39*
  Condensed Sans Serif *39*
    Drawing and Cutting *39*
    Lettering with a Flat Brush *41*
  Even-Stroke Sans Serif Roman *42*
    Uppercase Letters *42*
    Lowercase Letters *52*
  Sans Serif Roman with Different Stroke Widths *53*
    Uppercase Letters *53*
    Lowercase Letters and Numerals *57*
  Italic *58*
    Lowercase Letters *58*
    Uppercase Letters and Numerals *62*
    Alternate Method *62*
    Applications *62*

3 Advanced Course

Introduction *71*
Classical Roman Lettering *72*
  Roman Monumental Capitals *72*
  Rustica *82*
  Quadrata *84*
  Earlier and Later Roman Cursives *84*
  Uncials and Half Uncials *86*
  Carolingian Minuscule *88*
Gothic Letters and the Renaissance *90*
  Textura *90*
  Rotunda *909*
  Gothic Scripts and Variations *95*
    Gothic Cursive *95*
    Batarde *95*
    Schwabacher *98*
    Fraktur *98*
    Chancery Cursive and German Kurrent *105*
Renaissance Roman and Italic *106*
  Renaissance Roman *106*
  Humanist Italic *112*

From the Baroque to the Nineteenth Century *119*
   Baroque Roman *119*
   Neoclassical Roman *120*
   Seventeenth- to Nineteenth-Century Scripts *125*
   Nineteenth-Century Display Styles *130*
      Bold Roman Neoclassical *130*
      Egyptian and Italienne *130*
      Tuscan *130*
      Sans Serif *130*
Cyrillic *166*

4 Type and Lettering in Practice

Introduction *171*
Typography *172*
   Choosing Type *172*
   Mixing Type *173*
Calligraphy *174*
   General Remarks *174*
   Documents and Short Texts *175*
      General Remarks *175*
      Using Parchment *176*
      Raised Gilding *176*
   The Handlettered Book *179*
      General Remarks *179*
      Format and Page Layout *180*
      Designing the Text Block *183*
      Front Matter and Colophons *185*
      Planning the Layout *186*
   Binding Handlettered Books *187*
      Paper Structure *187*
      Folding Paper and Board *188*
      Folders, Covers, and Tubes *188*
      Binding Multiple Pages *188*
Lettering in Applied Graphic Art *190*
   Logotypes *190*
      Lettering in a Circle *192*
   Posters *192*
   Packaging and Labels *193*
   Book Jackets *194*
   Lettering for Exhibitions *194*
      Rubber Stamps *195*
      Stencils *196*
      Computer-Generated Lettering *196*
      Drawing Large Letters *196*
      Transfer Type *196*
Graphic Arts Tools and Procedures *196*
   Roughs and Finished Artwork *196*
   Sketching Type *197*
   Lettering on Photographs *197*
   Phototypesetting *198*
   Special Techniques *200*
      Woodcuts and Wood Engravings *200*
      Linoleum Cuts *203*
      Engravings *204*
      Resist Technique *204*
      Scratch Technique *205*
Lettering in Architecture and the Environment *205*
   General Remarks *205*
   Practical Hints *206*
   Special Techniqus *207*
      Neon Signs *207*
      Lettering on Walls *208*
      Lettering Design for Stone Carving *209*
      Lettering in Bronze *211*
      Lettering on Glass *211*
      Large-scale Lettering on Banners *211*

5 Portfolio of Type and Lettering in Practice *213*

Bibliography *249*
Index *251*

FOREWORD

Textbooks generally address members of a particular trade or profession. The *Handbook of Type and Lettering* does not. This book covers aspects of the design, use, and evaluation of lettering as it is practiced by graphic artists, advertising copy writers, sign painters, decorators, stonemasons, architects, typesetters, retouchers, bookbinders, engravers, art educators, teachers and students at colleges of fine art and related institutions, and interested amateurs. This book does not offer a theoretical discourse on the art of lettering; it does not claim to represent the current status of the art itself or the accomplishments of letterers in Germany or any other country. The purpose of this book is to encourage and guide the pursuit of beautiful and well-chosen lettering.

Today's professionals in the field rely almost exclusively on already existing and prefinished materials, such as graphically or photographically reproduced alphabets, stencils, rub-on type, stamps, and photocopies. Their creativity is limited to choosing, combining, and arranging type. This is not altogether a bad development. It is much better that the nonexpert should follow proven models rather than dabble in the creation of yet another amateurish new production. The use of original, newly drawn alphabets should be the domain of the adept. Type design and calligraphy play a constantly decreasing role in today's technologically controlled processes. Only a select few of all the artists who are involved in lettering have the opportunity to influence the design of type directly, yet for all of them familiarity with the areas of type design and application is indispensable and can only be achieved through a traditional course of studies in lettering and drawing formal alphabets. The object may be a sign, a logotype, or a poster; the design may have to stand on its own or be part of a larger idea; the aim may simply be readability or the transfer of emotion. A haphazard string of letters will not serve any of these purposes well. To control a situation is to take all elements into consideration. Even working with already existing material requires careful thought about the relationship of all contributing components and their positions on the page. The designer must be able to change images in a creative way according to the specific requirements of the situation, which is only possible if he or she has a solid understanding of the letters' anatomy. This understanding can be gained from practicing the art of lettering and by studying the forms of the letters.

The *Handbook* is a workbook that can be used in a self-study course by anyone who takes it upon him- or herself to follow the instructions. Since the text is often short on any given topic, it is important to read it in its entirety. Attention should be given to the introduction to each chapter. No book can be a substitute for a good teacher, but a solid foundation for the task of drawing and designing alphabets can certainly be gained from these pages.

The book offers a series of exercises for beginners (Chapters 1 and 2) and for advanced students (Chapter 3). Chapter 1 offers information on general design principles, Chapter 2 basic facts without which serious work with letters would be impossible. Step-by-step instructions are offered to compensate for the missing guidance of an actual teacher, but it is clear that self-study requires a great amount of discipline, since the student not only has to understand the critical phases of the learning process, but has to guide the process while being involved in it. Golwitzer likens this act to looking over your own shoulder.

As the learning process unfolds it will

take different directions depending on the learner's skill, talent, diligence, temperament, and creativity. No book can offer precisely programmed help on questions such as text format, choice of pens or techniques, or when to switch from work on simple forms to more complex problems.

Every student has to decide for himself where his abilities lie. It is much better to aim for a satisfactory performance on an "easy" project than to allow ambition to dictate an unobtainable goal.

The illustrations in Chapter 1 were chosen because they represent generally accepted forms. The author does not intend to use this text to peddle her own opinions, and the learner is warned against conscious and labored efforts to express his personality in his lettering style. "To learn what can be learned is not a hindrance to originality, but necessity and enrichment" (Hermann Hesse). Writing is the most important means of expression for human beings. It is more than just the visualizing of language, and precisely for that reason it is imperative that its form correspond to the job at hand. Craft, not far-fetched originality, is needed.

Last but not least I want to point out that this book does not teach what is commonly known as calligraphy. Rather it is intended to inspire an understanding of the forces of material forms and surfaces as a basis for good lettering, drawn or typographic.

The course for advanced students (Chapter 3) comprises the development of letters from Roman monumental capitals to the sans serifs of the nineteenth century. Assuming the mastery of skills taught in Chapter 2, I suggest that the lettering styles be studied in the order of their historic emergence, since each one of them built its form materials on the basis of earlier styles.

In Chapter 4 I have tried to offer illustrations that show as many different applications and inspirations as possible. Again, the beginner should be warned not to give in to the temptation to take on projects that are too advanced for his or her abilities.

I developed the manuscript with the help of the following texts: *Writing and Illuminating and Lettering* by Edward Johnston, *Deutsche Schriftkunst* by Albert Kapr, *Treasury of Alphabets and Lettering* by Jan Tschichold, and *Die schöne Schrift* by Frantisek Muzika. I am also indebted to the Schneidler writing school, and the Leipzig calligraphy tradition. I have utilized the skills and knowledge gained during my years of studying and teaching at the Hochschule für Grafik und Buchkunst in Leipzig.

Finally, I wish to express my gratitude to the late Mr. Heinz Braune, Mr. Harald Brödel, Mr. H.-J. Förster, Mr. Renate Herfurth, Mr. Karl-Georg Hirsch, Mr. Achim Jansong, Prof. Dr. Albert Kapr, Mr. Joachim Kölbel, Mr. Volker Küster, Mrs. Gera Kunzendorf, the late Mr. Fritz Przibilla, Prof. Walter Schiller, the late Mr. Kurt Stein, and Dr. Gerhard Winkler and the employees of the Schriftmuseum der Deutschen Bücherei and of the Bibliothek der Hochschule für Grafik und Buchkunst in Leipzig.

Hildegard Korger

CHAPTER 1

Advice on Design

Figure 1

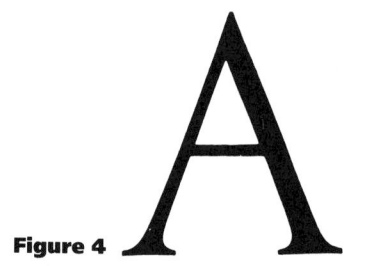

Figure 2

Figure 3

Figure 4

Figure 5

12

INTRODUCTION

Writing serves communication. It does so directly by making language visible, and indirectly by relating aesthetic values. Every letter and every functional unit of writing represents a sound of language and has its own visual value. Readability is the function of writing; its visual appearance is the form. The connection of form and readability constitutes design. "Good lettering requires as much skill as good painting or good sculpture.... The designer of letters, whether he be a sign painter, a graphic artist or in the service of a type foundry, participates just as creatively in shaping the style of his time as the architect or poet," wrote Jan Tschichold.[1] The range of possibilities in the Western art of writing reaches from ancient Roman inscriptions and Bodoni's *Manuale Tipografico* to the designs of Schneidler and Gaul, where words are mere pretexts for design opportunities; it includes the classical letterforms of Trajan's Column as well as the decorative and playful initials of Neudörffer or Frank, the clear and somber shapes of the sans serif, and the alphabet written by Karlgeorg Hoefer with bold strokes of the brush. The examples in Figures 1 through 6 and others in Chapter 4 of this book show that optimal readability is not the only possible aim of writing. Every specific situation, every artistic intention, requires that form and readability be balanced against each other. The laws of formalism exert themselves in the creation of single letters, in the relation of letters to each other and in their combination as a typeface.

These laws can be learned as a "grammar of design," in Walter Gropius's words,[2] but they are nothing more than the necessary prerequisites for the visualization of creative thought. Creativity itself is not based on the application of rules alone. To form and to shape would remain mere dexterity but never mature to an art if it were not for the specific aspects of individual talent, which—if present—can be developed and guided but never learned.

BASIC ELEMENTS OF DESIGN

Contrast

Every effect depends on contrasts, which appear either as pairs or alone. In the latter case an association is formed in our minds which completes the contrast that is assumed to be known.

In design we are always dealing with contrast pairs. Every measure needs a countermeasure to activate it. In relation to writing this means that the design unfolds on an empty page, the contours of the letters become visible only against the background of the white paper. Dark groups of letters draw their expressive value from juxtaposition with lighter ones. Large units look large only in relation to small ones, excitement needs repose, the effect of a color is influenced critically by adjacent colors.

Rhythm

The second element of design is rhythm. This may be created by regular repetition of like units or by repetition of similar units in opposition. If certain elements are stressed above the others, rhythmical values are increased. Spacing

1 *Roman monumental capitals. Commemorative inscription on black marble, found near St. Pantaleon in Cologne. (Photograph from the Römisch-Germanischen Museums, Cologne, from Albert Kapr,* Deutsche Schriftkunst.*)*

2 *A page from the* Manuale Tipografico *by Giambattista Bodoni. Parma, 1818.*

3 *A page from* Der Wassermann *by F.H. Ernst Schneidler.*

4 *The letter A in the style of a roman capital.*

5 *Decorative initial by Johann Neudörffer the Elder.*

6 *Initial by Karlgeorg Hoefer.*

---

1. Jan Tschichold, *Treasury of Alphabets and Lettering*. Reprint. New York: Design Press, 1992. Copyright © 1952, 1965 by Otto Maier Verlag, Ravensburg.

2. Walter Gropius, *Architektur* (Architecture). 2nd ed. Frankfurt/Main, Hamburg: Fischer Bücherei, 1959.

and proportion are of utmost importance, because they create tension and interest. If the units are similar to each other, repetition is strengthened and the rhythm is weakened. In extreme cases the rhythm disappears entirely. Design is therefore based on the rhythmical arrangement of subunits. It is based on tension, not on monotony. On a two-dimensional surface rhythm is realized through the repetition of graphic contrasts, or, more precisely, through contrast pairs. This can be accomplished in one of the following ways:

1. In the sequence of movements that are used to define the letterform — the elements of the rhythm and their relationship may be vertical, horizontal, diagonal, and circular.

2. Within the letterforms themselves there may be round or pointed, thick or thin, large or small parts. The interior and exterior of the forms may also contrast.

3. Through rhythmical structure created by such factors as the position of a form on the page and its relationship to the center axis; forms and interior forms of words, and the entire page layout itself; the structure of the text through paragraphing, sizes and weights of type; colors; the relationship of groups of words to empty space; the boundaries of the page and its height and width; empty space.

All these factors are interdependent. The rhythm of movement helps create the form. The rhythmical elements of separate shapes structure the page. Rhythm unites movement, form, and surface in an organic whole, and the same rules apply to small sections and to the entire work.

Size Relationships

We know from experience that the measurements that we perceive through our eyes are unreliable. Optical illusions have to be taken into account seriously. As Emil Ruder says, "The optical and aesthetic image in the mind is superior to the true geometrical construction; therefore black and white has to be balanced according to this mental image."[3]

Figures 7 through 16 illustrate some of the illusions we must deal with.

Figure 7 shows a geometrically correct square. It seems to be wider than it is high. Similarly, a true circle will appear to squat.

Figure 8 shows two rectangles of exactly the same size. The upright one appears to be narrower than the other one.

The two areas in Figure 9 are bordered by two horizontal and two vertical lines respectively. The former appears wider, the latter taller.

When horizontal lines are combined into a block as on a written page, the image seems taller (Figure 10).

The squares in Figure 11 are identical, but the one on the larger format seems smaller than the square on the smaller format.

Not only the immediate surrounding of a particular form or group of shapes is important, but also the sizes of neighboring forms. The middle squares in Figures 12 and 13 are again identical, but large surrounding shapes make one appear smaller than another.

Figure 14 shows that light-colored sections or letters of a negative type appear larger than darker spaces or letters of the same size. White or light colors radiate beyond their boundaries.

Figure 15 shows that a line with fewer segments will seem shorter than one of the same length with many segments. For this reason the lines of a wide or large typeface appear shorter than those of a smaller or narrower type.

Figure 16, finally, shows that tightly spaced vertical lines seem taller than widely spaced ones of equal measurements. Compressed or condensed letters will be optically taller than wider letters.

---

3. Emil Ruder, *Typographie. Ein Gestaltungsbuch* (Typography, a manual of design). Teufen A.R.: Verlag Arthur Niggli, 1967.

Figure 7
Figure 8
Figure 9
Figure 10

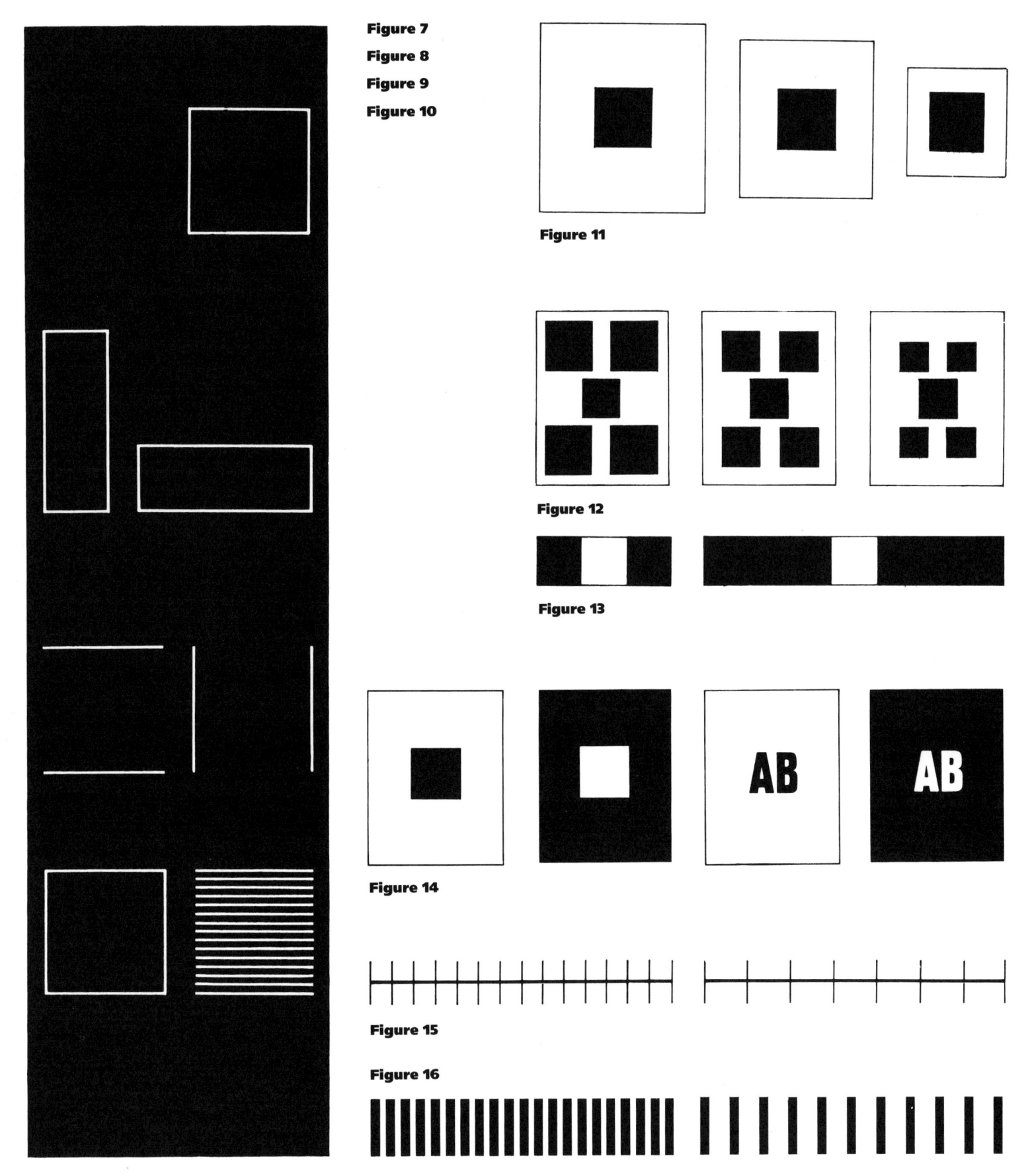

Figure 11

Figure 12

Figure 13

Figure 14

Figure 15

Figure 16

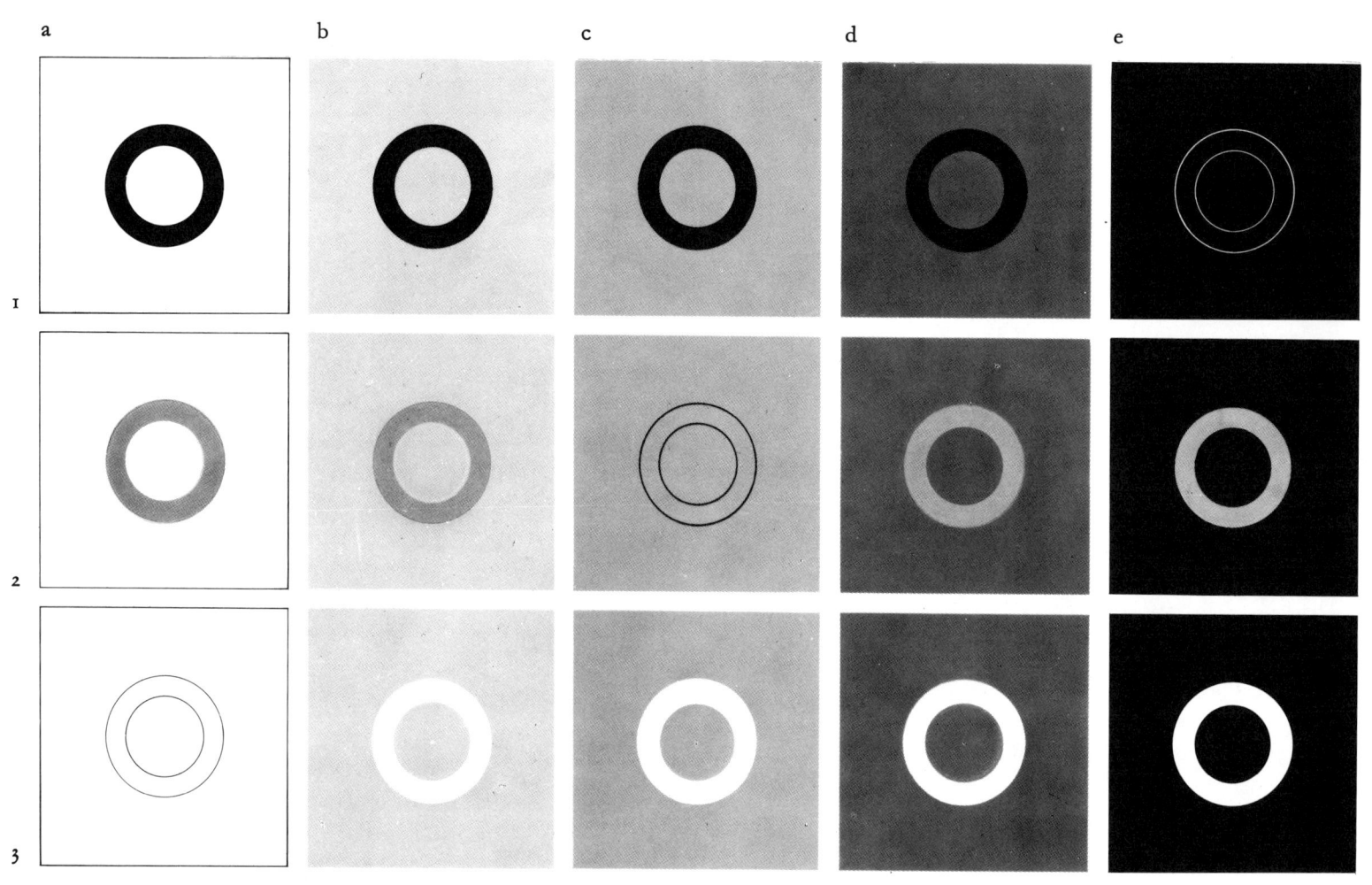

Figure 17

Color

In lettering color only serves and supports the aim of readability: it attracts attention and creates associations. A basic knowledge of color theory has to be assumed. A good text on the topic is recommended for further reference.

Light/Dark Contrast

Optical illusions have to be taken into account in relating different sizes to each other; a similar situation exists in color design. The values perceived by our eyes are rarely identical to the actual color values. Adjacent colors and light conditions influence our perception. Disregarding the strongest light/dark contrast that exists, between black and white, every hue of the color wheel has its own lightness or value, the lightest for yellow, the darkest for blue. Red and green have equal values. In addition, every color can be represented in lighter or darker shades. As seen in Figure 17, these gradations of value are of great importance for the relationship of lettering and background, since the legibility of the writing depends on them. The greatest contrast can be seen in fields 1a and 3e.

A pure white surface may be too harsh for the design of books, since its radiance causes the contours of the letters to appear fuzzy, and clarity is diminished. It is common, therefore, to use white paper with a slight warm tint for commercial applications. To prepare a wall for lettering, white paint should be mixed with some ocher or brown. Black can also be softened through the addition of brown.

In comparing the fields 1b, c, d and 3b, c, d in Figure 17, we can observe that the white circles seem more detached from the background than the black ones. This happens because white is a more active color than black and outshines the other colors. Thus, on colored backgrounds of medium darkness white lettering appears more clearly than black lettering. In row 1 the circle in field e seems even more prominent than those in fields b, c, and d because of its white contour. In row 2 it can be observed that the same gray circle appears darker on white, and lighter on black. Similar effects result when gray

16

circles are put on intensely colored backgrounds. The gray will shift towards the complementary color of the background — that is, it appears greenish on red, yellowish on blue. Another setup involves primary colors. If the gray circles are replaced by red ones on blue and yellow backgrounds, the red circle on the yellow will appear darker than the red circle on the blue, because it takes on some of the complementary color of yellow. This phenomenon is known as simultaneous contrast. The exception to it occurs when two complementary colors of the color wheel in their full strength are opposed. In this case both colors would appear enhanced. For optimal readability and visibility over greater distances the light/dark contrast of lettering and background is of utmost importance. A red and a green of equal value can cancel each other out visually and produce a flickering image.

Warm/Cool Contrast

No actual feelings of warmth and coldness are connected to colors. The words "warm" and "cool" represent "an association that the colors elicit in our minds."[4] The strongest warm/cool contrast exists between a yellowish red and a greenish blue. On the color wheel the warm colors are located between yellow and red, the cool ones between blue and green. The colors between red and blue are simultaneously warm and cool, the ones between yellow and green are neither warm nor cool. Warm colors seem active, attract attention, and advance from the surface; cool ones seem passive and recede into the surface.

Pure and "Broken" Colors

Every color loses some of its brightness when it is dulled or "broken" by the addition of gray, or, more effectively, by the addition of some of its complementary color. Pure colors appear prominent opposed to broken ones; the intensity of a pure color can be heightened considerably by juxtaposition with broken colors.

Color Associations

Colors have psychological and symbolic value, though associations vary depending on culture or even point of view. For example, the color green may conjure up the image of immaturity or, alternatively, environmental concerns. In some cultures white is associated with mourning, whereas in the West we generally think of it as pure, festive, cheerful, and noble. During the Middle Ages in Germany green, not red, was the color of love. These are just a few examples of different color associations. The associations also depend on whether the colors are light or dark, warm or cool, pure or broken, and are heavily influenced by what Albert Kapr calls "conditions of landscape, nationality, history, religion, and class."[5] Tradition and custom perpetuate them.

Colors also depend on the light under which they are seen. At dusk red appears darker, blue lighter. Artificial light alters almost all colors, and influences blue and green most strongly.

There are no recipes for good color combinations. Taste can be cultivated through the study of works of classical and modern painting or graphic art. Abstract works are particularly useful.

---

4. Paul Renner, *Ordnung und Harmonie der Farben* (Order and harmony of colors). Ravensburg: Otto Maier Verlag, 1947.

5. Albert Kapr, "Probleme der typografischen Kommunikation" (Problems in typographic communication), in *Beiträge zur Grafik und Buchgestaltung.* Leipzig: Hochschule für Grafik und Buchkunst, 1964.

A practical hint: cut out color samples from posters, wrappers, and packaging material or painted scraps and collect them in folders. Commercial color sample collections are a helpful tool. Colors that appear in print will differ from the ones used in design. A classification of colors can be made according to various systems; if your work will be reproduced, obtain color samples from the printer.

Movement and Focal Points

If several elements are present on one page, they cannot be comprehended at the same time. Movement is created as the eye follows the lines of composition in a quick and unconscious way. Figure 18 represents a pattern of basic movements that occur in many visual compositions. These movements are of course subdivided into secondary structures of more or less prominence. Presented here are only the physiological elements of normal movement, which are also present in handwriting.

In his book *Typographie*,[6] Emil Ruder lists categories of internal movement: increase or decrease of value or size, dissolution of compact elements and assembly of scattered elements into compact ones, movement toward the center and away from it, movement leading from the bottom up or from the top down, from left to right or from right to left, movement from the inside toward the outside or vice versa, along diagonals or following angular paths, and so on.

It is very important to realize that relationships of size and weight are influenced by the position of the elements on the page. If a space is separated along a horizontal line into two geometrically equal parts, the upper half will appear larger than the lower (Figure 20). All elements contained in the upper half will gain equally in prominence. This effect can be explained easily: in nature most objects in the upper half of our visual

---

6. Emil Ruder, *Typographie. Ein Gestaltungsbuch* (Typography, a manual of design). Teufen A.R.: Verlag Arthur Niggli, 1967.

18 *Adapted from Müller-Enskat,* Theorie und Praxis der Graphologie I *(Theory and practice of graphology). Rudolstadt, 1949.*

21 *From Max Burchartz,* Gleichnis der Harmonie *(Parable of harmony), 2nd edition. Munich, 1955.*

Figure 18

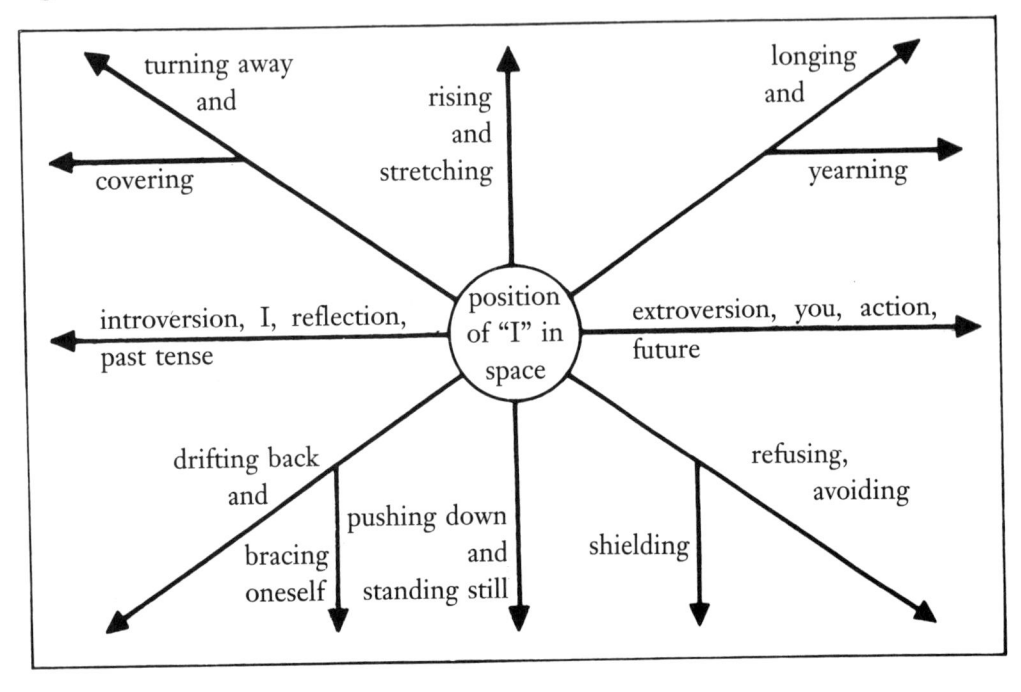

Figure 19

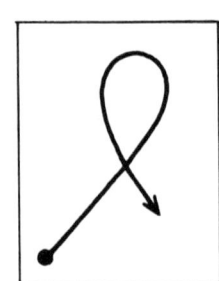

Figure 20

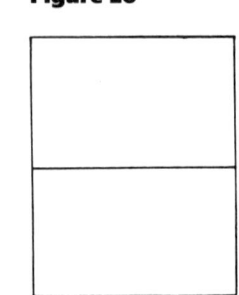

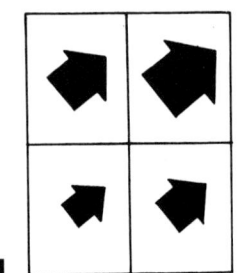

Figure 21

field are smaller, lighter, and brighter than the ones below. Big, heavy, or dark elements in an upper field therefore strike the observer as unusual and seem more active. This effect is particularly important in symmetrical arrangements, where the most important elements should be placed in the upper half. If the composition is asymmetrical, the relationship between the right and the left side is also important. The writing sequence of most occidental cultures moves primarily to the right, and psychologists draw a number of conclusions for the reading of movement and the interpretation of elements on a flat surface. The right-left contrast is interpreted as follows:

left	*right*
passive	active
introversion	extroversion
subject	object
I	you
past	future

Other associations can be made in reference to movements and their qualities within a composition. In his book *Point and Line to Plane*, Kandinsky says the following about the contrast of vertical and horizontal: "The horizontal line is reminiscent of the horizon, the ocean, the prairie, quiet, sleep, and death. It appears in a picture impressively, quietly and with importance.... When a vertical line rises in a picture, it appears uplifting and full of vigor, it reminds one of towers, obelisks, and fountains; sinking, it appears somber and heavy and reminds one of lead and heavy weight."

Tests run by advertising agencies found that elements of composition appear most prominent if they are arranged on the upper part of the surface rather than near the bottom. A position on the right side is more active than one on the left side. The bottom left area of a surface offers the least impact (Figure 21).

These concepts can only be of use if they are not considered ironclad rules. No beginner should be tempted to place all important design elements in the upper part of the piece as a matter of course. Such a constant repetition would be dull indeed. "Artistic means can lose interest quickly; a change is needed for a fresh impact." This is the teaching of Ernst Schneidler in his work on design, *Der Wassermann*.[7] The information that has been given about movement on the surface and the positioning of elements should merely make it easier to choose the size and weight of these elements so that an optimal composition can be achieved.

---

7. Ernst Schneidler, *Der Wassermann. Ein Jahrbuch fur Buchermacher*. Stuttgart: Akademie der bildenden Kunste, 1948.

Three-dimensionality on a Flat Surface

Design elements can create a three-dimensional effect on a two-dimensional surface by their sizes, relationships to each other, and their absolute position.

Big and heavy elements seem to protrude from the surface—they jut out towards the observer—while smaller and lighter forms recede. Figure 12 shows an example of this effect. In the presence of vertical lines, horizontals recede into the background. Irregular surfaces prominently advance in front of flat ones. Similar effects are well known in elements of different hues (see the section on Color Associations, page 17).

Three-dimensional effects can be counteracted by judicious use of these principles. An element that seems to protrude because of its own properties can be placed in an area that relegates it to the background; a small, receding element can be emphasized by placing it in a prominent position.

Figure 22

Figure 23

HAEOMmz

OPTICAL ILLUSIONS IN LETTERING

A good eye is much more important than measuring equipment when the aim is a well-drawn or -written letter.

Figure 22 shows rectangles, triangles, and circles of the same height. Nevertheless, the circles and triangles seem smaller than the rectangles—the circles because they touch the guidelines only at two points; the triangles because they only touch at one point, at the tip. Unless the arc and the tip extend slightly beyond the guidelines, the objects "fall through" (Figure 23).

In Figure 24 the circle was placed exactly in the middle of the rectangle, the horizontal line exactly in the middle of the vertical. Neither element, however, seems to occupy the center position: the upper half seems larger. To counteract this effect, the separating elements are slightly moved up in Figure 25. They now occupy the "optical center." The crossbars of letters such as H, E, B, K, and S should not be placed in the geometric, but in the optical, center position. The upper part of these letters should be smaller than the lower part. In round letters such as O, D, G, and C, the upper part should be more expansive to give the letter a certain tension (Figure 26).

A vertical separation makes an area appear narrower; a horizontal separation makes it seem wider (Figure 27). Serifs have a similar effect on the proportions of type: they stretch or widen the form (Figure 28).

A bar appears long and narrow if it is in a vertical position but wide and heavy as a horizontal element (Figure 29). This effect is important in the construction of sans serif and Egyptian-style letters. If the horizontal elements of a letter are to look like vertical ones, they have to be drawn somewhat narrower (Figure 30). The same rule applies to round parts: they have to be slimmer at the top and bottom areas, while at the sides they have to be considerably wider than the straight elements. Diagonals require a special adjustment; here the angle is of importance. Figure 31 offers guidelines for the construction of sans serif letters.

Even vertical lines require differentiated stroke widths (Figure 32). The juxtaposition of E and B shows that the

30 *Stroke widths of the capital letters of Adrian Frutiger's Univers. From* typographia *9/65.*

Figure 24

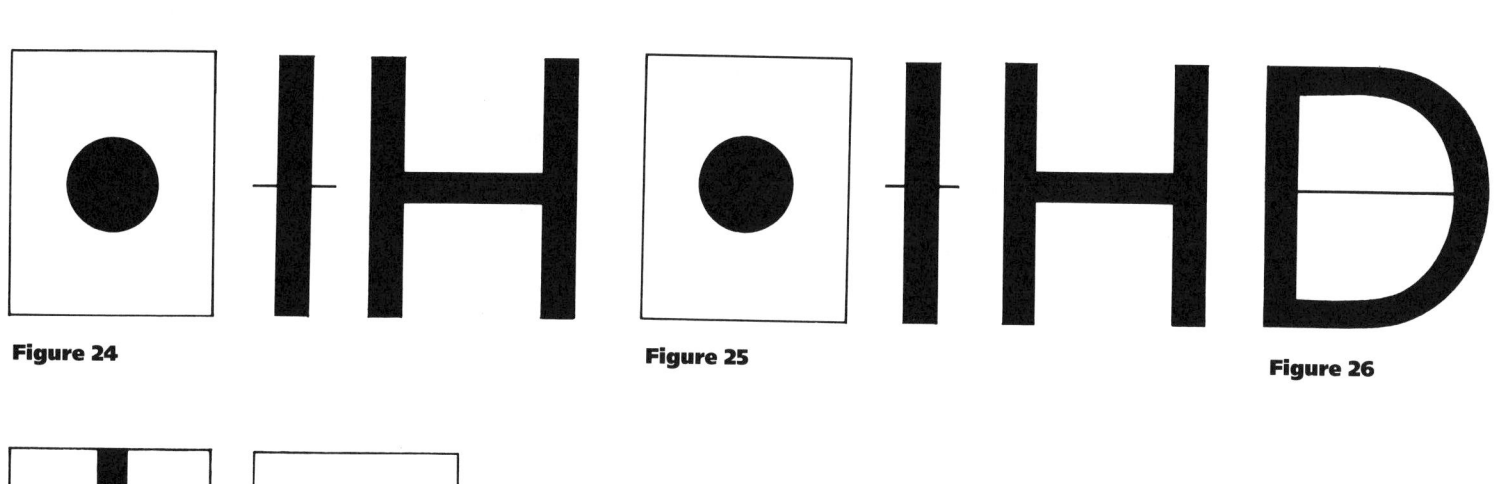

Figure 25

Figure 26

Figure 27

Figure 28

Figure 29

Figure 30

Figure 31

Figure 32

letter B appears darker than the letter E. An experienced letterer will narrow the vertical lines of the B and slightly widen those of L or I to balance. Follow the same principle for roman letters.

Figure 33 shows that a short vertical bar appears thicker than a longer one of the same width, which is why it is necessary to draw short verical parts of letters with slightly thinner strokes. For the same reason, all lowercase letters should be constructed with thinner verticals than uppercase letters (Figure 34). The larger interior space of uppercase letters also necessitates different stroke widths.

When two diagonal strokes intersect, a black spot results and draws undue attention. Counteract this effect by deeper cuts or angles at these points (Figure 35).

The transition from straight to round lines also presents problems. The careful observer can detect a break in the line if the round section was constructed with compasses. Sacrifice the perfect geometric form to a gradual change in curvature (Figure 36).

If you construct condensed sans serif capitals such as O, C, G, and D with compasses and ruler, the straight lines will appear slightly concave. To make

Figure 33a

Figure 33b

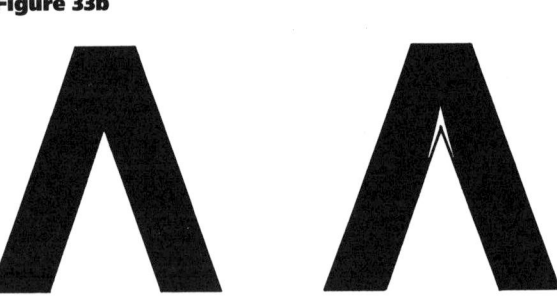

Figure 34

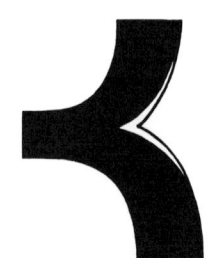

Figure 35

Figure 36

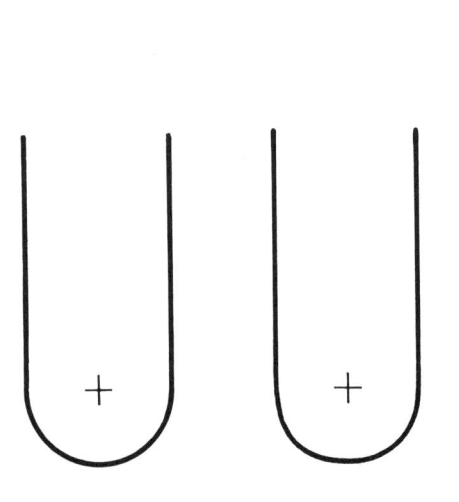

Figure 37

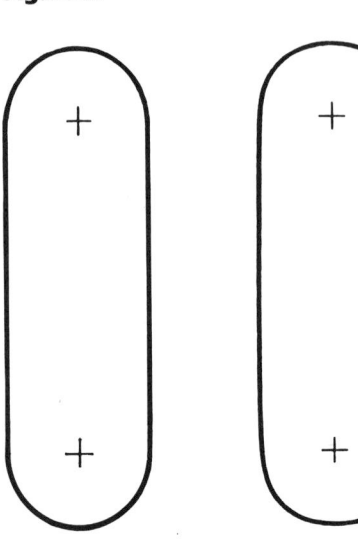

Figure 38

Figure 39

Figure 40

Figure 41

WOLLGARNE

WOLLGARNE

Figure 42

WOLLGARNE

Figure 43

WOLLGARNE

MOHN **MOHN**

Figure 44

Figure 45

mohn mohn

mohn mohn

them look straight, draw them with a slight bulge towards the outside (Figure 37). Similarly, letter parts that are geometrically perfectly straight will seem thicker in the middle, but if you narrow the verticals in the midsection, the sides will appear parallel (Figure 38). The smaller a form is, the rounder its corners will seem (Figure 39).

LETTER, WORD, AND LINE SPACING

Letter Spacing

Letters are arranged next to each other as words and lines, lines are combined to make text blocks, and all are elements of the page layout. It would satisfy neither the demands of legibility nor of aesthetics to string one letter after another without further thought. Empty spaces inside the letters and the area surrounding them all have specific ornamental values, and the designer has to deal with them (Figures 40 and 41). Spaces can be neutralized or activated through manipulation; more information on this can be found at the end of this chapter.

It is necessary to render the interior spaces of letters optically neutral to achieve clarity. If all spaces between letters were exactly equal (Figure 42) or if they were decided in an arbitrary way, areas of concentrated color would appear on the page. Letters such as O, C, G, on the other hand, would create holes, and the surrounding spaces of L, T, A, V, P, and others would act in a similar way. The result would be unba-

42 *Letters with equal spaces between them.*
43 *In a well-balanced word the spaces between letters are optically balanced.*

23

lanced, ugly, and difficult to read—it might even be necessary to read certain words letter by letter. To preserve the unity of each word, the spaces between the letters need to be balanced optically; no single form should dominate the others (Figure 43). Two neighboring vertical lines need more space between them than a vertical next to a round shape or a diagonal. Consider both the space between letters and the space surrounding the letters in question to find the optimal distance. Letter spaces seem to intrude into interior spaces of letters that have open sides, and the effect is additive.

No absolute values can be given for letter spacing. As a rule a line can be considered balanced when the letter O, placed between vertical lines, does not seem to fall out of the text. Practice with the group of letters MOHN (Figure 44).

For lowercase letters use the equivalent of the inner width of the letter m as a guide for the distance between letters. If the inner width is extreme, or for sans serif letters, make the spacing somewhat narrower (Figure 45).

The easiest way to achieve a balanced layout is to construct all letters of the word or line separately, cut them out, and glue them down in the new arrangement for photographic reproduction. Start with the largest necessary space between two letters and decide on the other spaces accordingly. The space between L and A is often the widest one. Do not try to solve this problem by shortening the crossbar of L or the serifs, or by squeezing the L too closely to its neighbor. Keep in mind that a necessarily wide spacing between L and A is aesthetically preferable to maimed letters. If you write a word or a line in its entirety, you can correct the spacing either when you copy the sketch onto tracing paper or onto the final page.

Train your hand and eye through frequent exercises. Start with groups of letters like MOHN or mohn, be patient, and you will acquire a feeling for the right spacing of your letters.

If this aim seems elusive, write your word over and over again in the form of a list and make adjustments to the spacing until you are pleased with the result. Another way would be to cut the word apart and correct the spacing, but do not overlook the fact that this process produces shadows around the edges of the cut letters. Use the correctly spaced word as guide for later sequences that contain diagonals and round forms.

For emphasis the letter spaces can sometimes be increased beyond what is necessary for balance. While capitals may be set extra wide, lowercase letters may not, but since legibility is at stake, this maneuver should be used sparingly and with selected words only.

If the letters are spaced very tightly, their interior forms become the prominent design feature, decorative aspects become primary, and legibility suffers. Posters, book jackets, and similar material often utilize these factors (Figure 47). In other cases letters are spaced so widely that they are spread across the page in a purely decorative manner. Even though ease of reading may be compromised, the surprise effect of these arrangements can attract the reader's attention and is a legitimate tool that has many creative possibilities. It is self-evident that frequent repetition diminishes this effect, and the beginner especially is warned against copying such layouts.

47 *The ornamental value of the letters' interior spaces can be stressed when the letters are very close to each other. (From an advertisement of the Schmidt Brothers ink factory. Design by Olaf Leu.)*

AUSGLEICHEN
SPERREN

Figure 46

Figure 47

Word Spacing

Keep spaces between words just as wide as necessary for clarity. The line should flow like a ribbon of even gray; large spaces between words rip white holes in it. If the line is made up of capitals only, choose a word distance that is equivalent to the width of I plus the necessary letter space on each side. A good space between words in lowercase letters is approximately equal to the x-height, unless a letter with round forms (o) or diagonals (v) stands at the beginning or the end of a word, in which case the space needs to be slightly reduced.

Line Spacing

Important factors in line spacing are: letter size, stroke width, density of the layout, and length of the lines. The letter size, in turn, is determined by the purpose of the text and its intended effect. Stroke width and density depend on the chosen lettering style. Roman appears lighter and needs a wider space between lines than textura, fraktur, or condensed sans serif, all of which appear heavy and call for closer line spacing. Lines that are made up of capitals only need less space between each other than lines of lowercase letters.

As a rule longer lines require more distance from each other than short ones, and lines of small letters seem long compared to lines of equal length that are made up of bigger letters (see Figure 16).

Consider the importance of legibility in decorative texts. Short lines make frequent word hyphenation necessary, and complicate the reading process, but long

48 *Roman. There is too little space between the lines. The words look like brambles and are hard to read.*

Schrift kann zur Kunst werden – aber zuvor einmal ist sie Handwerk. Zur Kunst wird sie nur dem, der das Handwerk beherrscht. Der rechte Schreiber gibt auf seiner jeweiligen Stufe das Bestmögliche, und diese Einstellung zur Arbeit ist ihm Quell ständiger Genugtuung und Arbeitsfreude. Eine besondere Verant-

49 *Roman. There is too much space between the lines. The text cannot be read smoothly, and the intervals appear more important than the lines themselves. Space restrictions limit the number of lines in this sample, but the effect would be the same even with three times as many lines.*

Schrift kann zur Kunst werden – aber zuvor einmal ist sie Handwerk. Zur Kunst

wird sie nur dem, der das Handwerk beherrscht. Der rechte Schreiber gibt auf

seiner jeweiligen Stufe das Bestmögliche, und diese Einstellung zur Arbeit ist

ihm Quell ständiger Genugtuung und Arbeitsfreude. Eine besondere Verant-

50 *Roman. Just the right spacing between lines makes the text easily legible and produces an even value of gray for the entire text block.*

Schrift kann zur Kunst werden – aber zuvor einmal ist sie Handwerk. Zur Kunst wird sie nur dem, der das Handwerk beherrscht. Der rechte Schreiber gibt auf seiner jeweiligen Stufe das Bestmögliche, und diese Einstellung zur Arbeit ist ihm Quell ständiger Genugtuung und Arbeitsfreude. Eine besondere Verant-

51 *Because the condensed sans serif has a relatively dark appearance, it is possible to arrange the lines in a closely spaced pattern.*

Schrift kann zur Kunst werden — aber zuvor einmal ist sie Handwerk. Zur Kunst wird sie nur dem, der das Handwerk beherrscht. Der rechte Schreiber gibt auf seiner jeweiligen Stufe das Bestmögliche, und diese Einstellung zur Arbeit ist ihm Quell ständiger Genugtuung und Arbeitsfreude. Eine besondere Verantwortung obliegt dem Gestalter von Schriften, die reproduziert

lines are difficult to follow and may tire the reader. In an English text eight to ten words or forty to fifty characters per line seem to produce the right length for one line. Keep the lines separated enough to preserve the ribbon effect, but close enough to achieve an even value of gray for the entire text. Consider both horizontal and vertical dimensions of the text block and the relation of its gray effect to the rest of the page. Open line spacing calls for wide margins at the top and the bottom of the page, close line spacing can be balanced with narrow margins. The examples in Figures 48 to 51 show how important the correct line spacing is for the legibility of a text.

To find just the right spacing, letter several lines, cut them apart, and move them around on paper until you like what you see. (The edges of the cut strips may cause irritating shadow lines.) Sometimes a minute variation is all that is needed. A good way to educate one's taste is to study the works of masters.

COMPOSITION

The aim of composition is to create harmony among elements that are in opposition to each other. Such opposites can be empty/full, whole/fragmented, light/dark, or different sizes and positions. Empty parts of the page and the inner areas of letters are more than just background; they are as important as the letters themselves. These negative spaces are design elements in their own right. The contrast between the letters and the surrounding space is not only necessary; it is crucial.

Principles of Composition

Never try to force a text into a preexisting compositional scheme. Every project deserves a fresh start according to its purpose and the materials at hand.

Layouts can be symmetrical or asymmetrical. Symmetry causes a static, monumental, and ceremonial appearance, and can even be stiff and artificial. All these impressions are triggered by the all-important central axis of a symmetrical arrangement, which contradicts not only the natural text flow from left to right but also the asymmetrical anatomy of most letters, which is the result of this movement. To place a line exactly in the middle of a page requires a great deal of work. Try to give weight to the upper part of the text block and avoid any appearance of three-dimensionality that could result from different line lengths.

It is often very difficult to combine the optimal design of a document or inscription with a logical text distribution. If words have to be hyphenated or if optical emphasis falls on unimportant words, it is best to abandon symmetry and opt for a flush-left arrangement. Even titles and headings may resist symmetry.

Asymmetrical arrangements result from organic movement. They are richer in contrasts and therefore more expressive. Experience in dealing with problems of spacing, sequence, and rhythm are essential for successful composition.

52   *Sketch for placement of an initial capital.*

53   *Sketch for placement of an initial capital.*

54   *Sketch for placement of an initial capital.*

55   *Page with handlettered initial capital from the illustrated book,* Peter Vischer. *Dresden: Verlag der Kunst, 1969. Designed by Horst Schuster.*

56   *Page with handlettered initial capital by Imre Reiner. From Imre and Hedwig Reiner,* Lettering in Book Art. *St. Gall, 1948.*

57   *Page with woodcut initial capital by Gert Wunderlich. From the illustrated volume* Sachsenhausen, *Berlin: Kongress Verlag, 1962.*

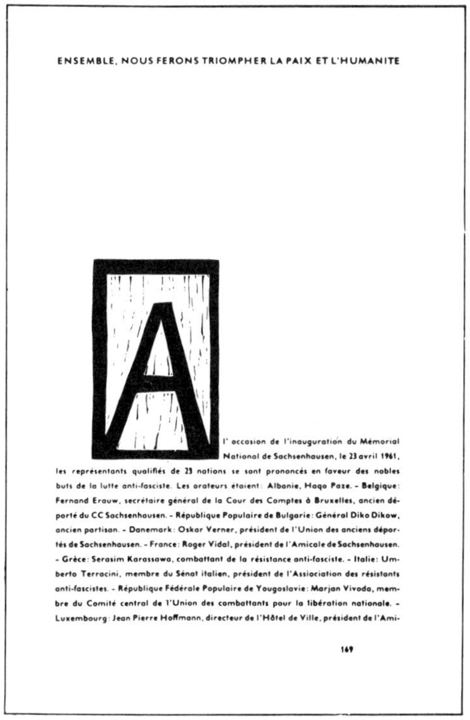

27

Layout

Layout is the process of arranging the text on the page. Text can be laid out in different ways. The arrangement, which makes the text readable and creates formal qualities, can be achieved by spacing, by emphasizing certain groups of letters or words, or by suppressing the effect of others. Different options are available, and should be chosen according to the context and aesthetic requirements of the text, never because of formalistic considerations alone. Avoid monotony, but be aware that frequent repetition of the same accents can be equally boring. "The means have to justify the end," Bertolt Brecht said. Be decisive about the use of your elements; halfhearted decisions dilute the effect of the whole.

A common procedure to create emphasis is to use letters of different sizes or weights. Single lines can be executed in full-size capitals or in small capitals. In longer texts it is better to emphasize sections by using italic, but, to say it again, lowercase letters should not be spaced extra-wide under most circumstances.

Decorative initial capitals, often several times the height of other capitals, are a favorite way of calling attention to the beginning of a chapter or paragraph. Another is to stress headings. In both cases it is important to find the right relationship in weight and size. Letter initials in different sizes, colors, styles, and characters, cut them out, and see which one best serves the purpose. Figures 52 to 57 show several possible arrangements of initial capital and text block.

It is also important to maintain a connection between the initial and the rest of the word of which it is a part. If the initial is to be freestanding, provide a connection to the first line of the text; if it to be incorporated in the middle of the text, connect it in some way to the line that it logically belongs to. If the initial is inscribed in a rectangle, align the horizontal stroke at the bottom with the baseline of the text. The flourishes of ornamental initials may extend into the left margin. Headings are usually conspicuous enough if they are separated from the body of the text by one or more line spaces. If extra emphasis is desired, use the italic that corresponds to the text lettering or any of the other means that have been described earlier.

Many variations in the gray values of a text can be achieved by adjusting sizes or weights of the letters and by the spaces between letters, words, and lines. The introduction of color adds another important element. Colored letters form a contrast to black ones and to the gray that results from the mass of the black letter. If hues are used sparingly, they create a jewel-like effect, but a color element can disappear in an overwhelming text block if it is not strong enough. Too many small elements of color spread over the text create a scattered effect—the eye is unable to distinguish among the weak contrast pairs and registers an evenly mixed color.

It is not advisable to use more than one additional color. Frequent choices are vermilion with the addition of some carmine red, a greenish blue, or gold, mixed from yellow with a touch of red and blue. Other options are a dark red or gray with some yellow or green. The addition of a little white to these colors will prevent the formation of spots when they dry, and it will keep them in the desired shade of lightness. The mixing should not result in dirty-looking or chalky colors. Mixing different type styles is discussed at the beginning of Chapter 4 and the sections on handwritten books also deals with ways of highlighting text.

Figure 58

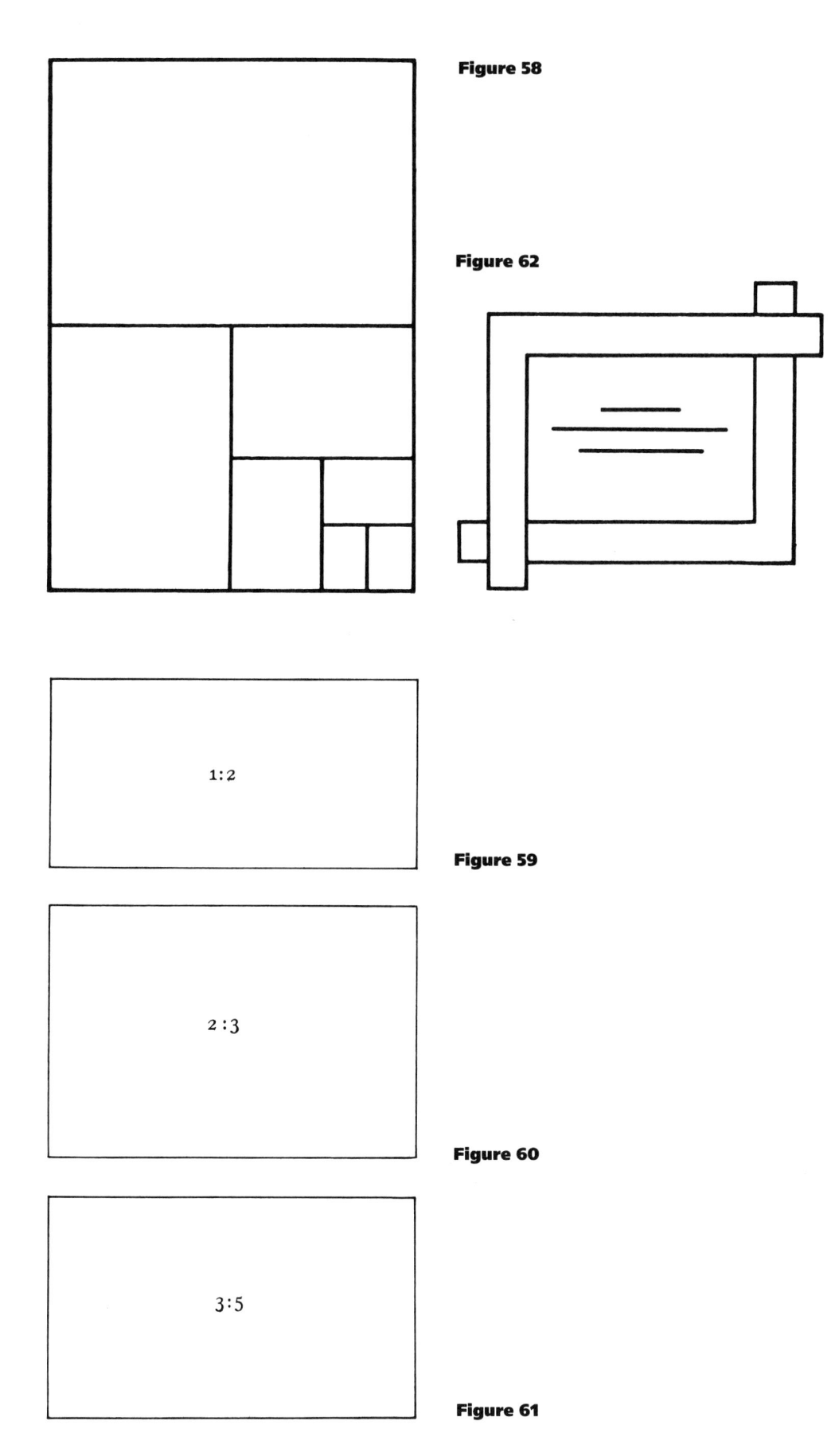

Figure 62

1:2

Figure 59

2:3

Figure 60

3:5

Figure 61

FORMATS

The graphic artist rarely chooses the format of his or her work: it is usually necessary to find a balance between text and a predetermined format. Letters, brochures, labels, and posters have to conform to specific criteria when it comes to size. In the metric international paper size system, an initial sheet (A0) is halved repeatedly (Figure 58), resulting in a sequence of ever smaller sheets. The digit after the A indicates how many times the original sheet has been cut. The most common sizes, with their equivalents in inches, are:

	Millimeters	Inches
A0	841 × 1189	33.11 × 46.81
1	594 × 841	23.39 × 33.11
2	420 × 594	16.54 × 23.39
3	297 × 420	11.69 × 16.54
4	210 × 297	8.27 × 11.69
5	148 × 210	5.83 × 8.27
6	105 × 148	4.13 × 5.83

In the United States the standard sizes of paper vary depending on the type — bond (writing), book, board, label, etc. Some common sizes are (in inches):

8½ × 11	19 × 24
9 × 12	19 × 25
11 × 14	20 × 40
14 × 17	22 × 30
15 × 22	25 × 38

The format of brochures often can be manipulated through different folds. The technical restrictions of the given standard sheet sizes influence book formats, but many variations can be encountered.

Jan Tschichold suggests simple ratios of length and width such as 1:2, 2:3, 3:5, or the golden section, 21:34. Labels look good in a ratio of 1:3.[8]

Sometimes it becomes necessary during the course of a project to change the format. A frame consisting of two L-shaped pieces of cardboard is of great help, especially if they are of a color that contrasts with that of the paper. Move the parts of the frame around until text block and remaining margins are balanced (Figure 62).

---

8. Jan Tschichold, *Treasury of Alphabets and Lettering*. Reprint. New York: Design Press, 1992. Copyright © 1952, 1965 by Otto Maier Verlag, Ravensburg.

PUNCTUATION MARKS

Every text designer or typographer has to be well versed in the use of punctuation marks. Literature on the subject is easily available and should be consulted. Here we touch only on the most frequently used items. Set periods, commas, colons, semicolons, apostrophes, and hyphens immediately after the preceding word. Quotation marks and parentheses also precede and follow the word they enclose without additional space. French quotation marks, or guillemets (» «), are preferable to German ones („ "), because they point towards the quote. Hyphens used for word breaks, should not be confused with the longer dashes (em, en), each of which has its own uses. Fraktur uses a double hyphen (=). If words have to be divided, check the dictionary or follow the relevant grammatical rules and apply them according to aesthetic standards. Two letters of a word do not look good if they stand all alone in a line. Avoid a repetition of more than three word breaks at the end of consecutive lines. It is helpful to set short lines ragged right, to reduce the need for hyphenation.

Check the dictionary for the correct spelling of compound words: some require hyphens, some are spelled open, and some closed.

Use the ampersand only where required by a company name.

63 *Study by Friedrun Weissbarth.*

64 *Letter elements in a composition for the back cover of a catalog. Designed by Christian Chruxin.*

CHAPTER 2

Introductory Course

INTRODUCTION

"Art takes time, and those who wish for quick results had better not start. Good lettering is, as a rule, created slowly. Only the master can, on occasion, work quickly; that is what makes him a master. Lettering, like all art, is not for the impatient."

These words of the famous typographer Jan Tschichold[1] head this chapter to emphasize that success is achieved only by a consistent and patient student. One or two repetitions of the recommended exercises will not suffice. They have to be done over and over again; in addition, if you are a beginner, you should devise new drills for every alphabet until you master all forms with ease. It would be detrimental to drop a difficult or boring section of the introductory course to get to the next exercise or a new alphabet. A constantly changing point of view does not foster a solid understanding of shape and form. Basic forms have to be mastered before derivative forms can be tackled. Only this sequence makes the learning process simple.

Mastering the basic forms, however, should not be the student's only goal. Of equal importance is the layout of the entire page, requiring coordination of many elements of graphic design. Simple exercise pages, trial applications, calligraphy, decorative arrangements of letters, monograms, symbols, labels, designs for posters, book covers, record sleeves, and many other projects should be approached in this way. Every unit should be followed by appropriate applications of the newly learned material.

---

1. Jan Tschichold, *Treasury of Alphabets and Lettering*. Reprint. New York: Design Press, 1992. Copyright © 1952, 1965 by Otto Maier Verlag, Ravensburg.

65 *Renaissance scribe.*

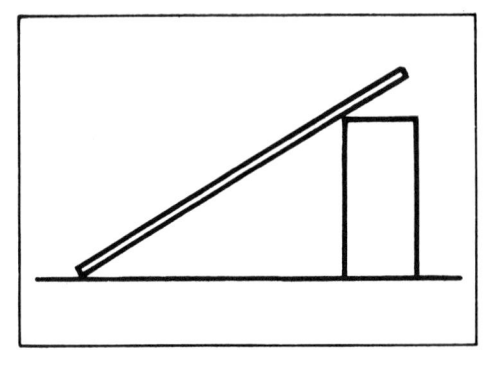

Figure 66

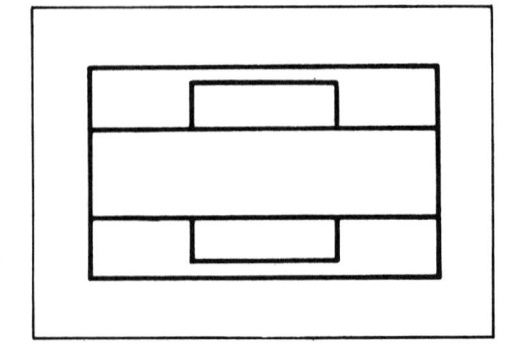

Figure 67

WORK SPACE, MATERIALS, AND TOOLS

The Work Space

Medieval scribes worked at desks with a slanted surface, which allowed an upright and unencumbered posture. Such an arrangement also affords a better view of the page and lets the ink flow more slowly from the nib. The desirable angle of 30 to 40 degrees can be achieved by fixing a board in the right position, as shown in Figure 66, or by tilting the entire table. For drawing, a smaller angle can be used. The work area should be about 24 by 36 inches (594 by 841 millimeters), large enough to support both elbows.

Only while guidelines are being drawn should the paper be fixed to the board; while you are lettering it has to be movable. Stretch a strip of paper across the board as protection from your hand, position the loose writing paper underneath and move it up as each line is completed (Figure 67). You will choose the height of your letters by experience. An inkwell and a brush to fill the nib are positioned at the left side of the drawing board. The left hand holds the brush to fill the nib of the pen, which is moved to the left side and remains in the right hand. The light source should be at the left side or in front of the writer. (Reverse these directions if you are left-handed.) Tools in good condition and a neat work space are essential for success.

Paper

You will need good-quality paper that is woodfree, will not let ink bleed through, contains the right amount of size, and is not too smooth. If the pen cannot be moved freely across the surface, the paper is not smooth enough. Do not use tracing paper. Among the best choices are book printing paper or Ingres paper. The popular watercolor papers are not suitable for writing, because their surfaces are too rough. The best paper size for practice sheets is about 11 by 14 inches (297 by 420 millimeters).

For drawing projects use a sturdier white, smooth, woodfree paper that can withstand the abrasion of erasers.

Never roll up either unused sheets of paper or finished work for safekeeping. It is best to keep them in folders and store them horizontally.

Ink

Use ink that covers the paper well, flows easily from the nib, and does not clog it. For beginning exercises use black ink only, because it makes errors more easily visible than lighter shades or other colors. Many commercial liquid inks are available: be sure the brand you choose will not corrode your pen nib, clog it, or produce fuzzy and imprecise lines. The best choice is a solid block of Chinese ink, which has to be prepared with water on a slab of slate before it can be used. Once dried, this ink crumbles and cannot be used again. Prepare small amounts and replenish the supply frequently. Black watercolor may rub off, but can be used if water-soluble glue is added. If you add prepared Chinese ink to watercolor you will get a very deep velvety tone of black. Add small amounts of ocher or red to your ink instead of using pure ink for any project other than exercises.

Draw with tempera or gouache colors and add water-soluble glue if necessary. Small and delicate embellishments are best drawn with Chinese ink, since it is composed of particles that are more finely ground than other pigments, and it can be spread onto the paper in a thinner layer.

Mix the ink in a small bowl until it has the right consistency and flows off the brush easily. Cover the bowl when it is not in use to keep it clean.

Pens and Nibs

If you want strokes of equal thickness, choose a pen with a round-ended nib like the one shown in Figure 68. Hold the pen at a 45-degree angle while making straight and round strokes, and keep the entire flat part of the nib on the paper. For letters with variable strokes use an oval or flat-edged nib (Figure 78). Some strokes require changing angles or even twists of the pen. If the width of the stroke exceeds ⅛ inch (3 to 4 millimeters), choose a broad flat-edged nib or a bamboo pen.

Nibs cut from quills and reeds make much more pleasing writing instruments than those made from steel, and the letters they produce seem warmer and sturdier. Another advantage is that they can be made at home in any desired width and with a variety of angles. The procedure is simple: take a thin reed or bamboo stick about 6 to 8 inches (15 to 20 centimeters) long, and cut a profile following the illustration in Figure 72. A scalpel with one sharpened edge is the best tool for cutting. If the reed has one flat side, use that for the writing edge of the nib. To give the nib an even width, the other side has to be shaped too. Scrape out the soft inside and determine the width of the nib with a cut on each side, as shown in Figure 73. In the middle of the nib make a cut about 1 inch (2 centimeters) long. Very wide nibs might require two or more such cuts.

To ensure that the cut will not split further during use, drill a small hole at the end of it with a bit or a hot needle.

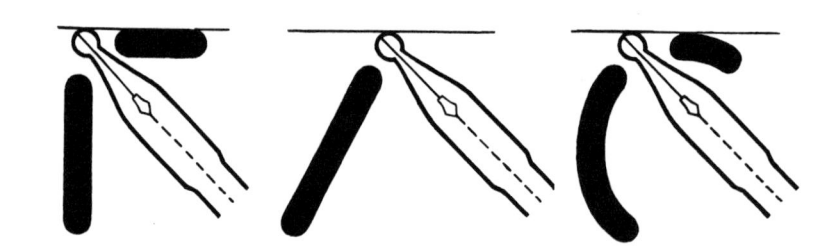

Figure 68

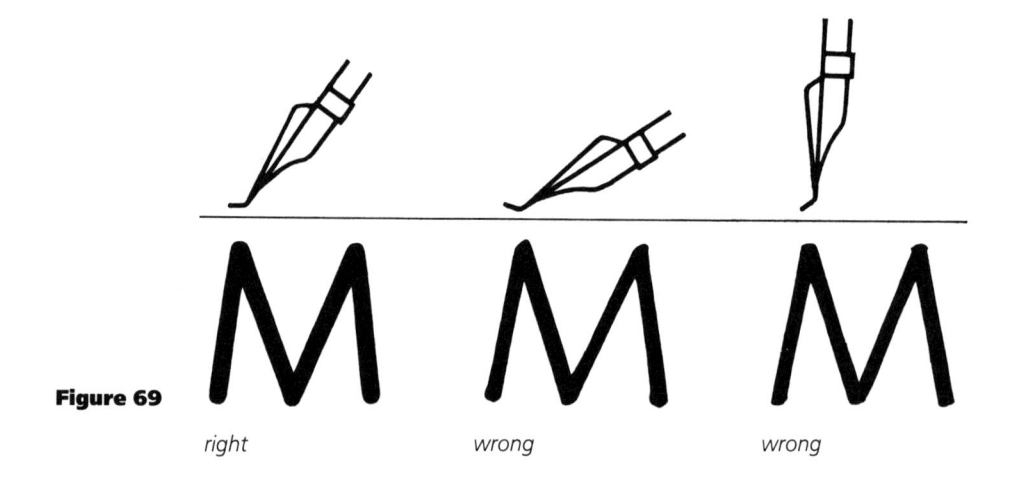

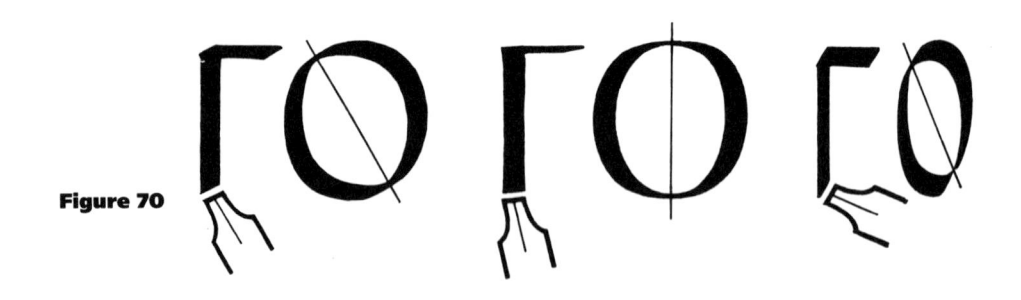

Figure 69

right   wrong   wrong

Figure 70

71 *Pressure distribution in writing with a pointed nib.*

upstroke   downstroke

Finally, place the pen on a plate glass surface and cut the writing edge once more with a knife. A slanted cut removes the front end of the slit (see Figure 74). The slant of the writing edge can be matched to the specific requirements of any writing style. Right angles as well as right or left slants are possible.

Quill pens can be prepared in a similar way. Take a sturdy turkey feather and remove the fine hairs on the shaft. Start the slit with a razor blade and force it open to a length of ½ inch (1 to 1.5 centimeters) with the aid of a brush handle. Should the slit retain a slight gap, bend the two halves of the tip slightly downward. The tips have to be even, and the edge should be slanted like that of a reed pen.

An ink reservoir does just what its name suggests. In reed pens use a 2-inch (5-centimeter) long strip of pliable tin, cut to the width of the nib, and insert it into the pen until the edge of the metal strip is about 1/16 inch (2 to 3 millimeters) behind the writing edge (Figure 76). In steel nibs the distance can be even smaller, but in no case should the strip interfere with the writing process.

Some steel nibs have ink reservoirs attached on top. They should rest lightly on the nib to facilitate a steady flow of ink. Fill them only halfway and make some trial strokes to use up any excess ink before you use the pen in your work.

Pointed nibs should only be used for scripts such as English roundhand or for drawing.

Clean your nibs periodically during long work sessions, because the writing liquid evaporates slowly and leaves a residue that can clog the tip.

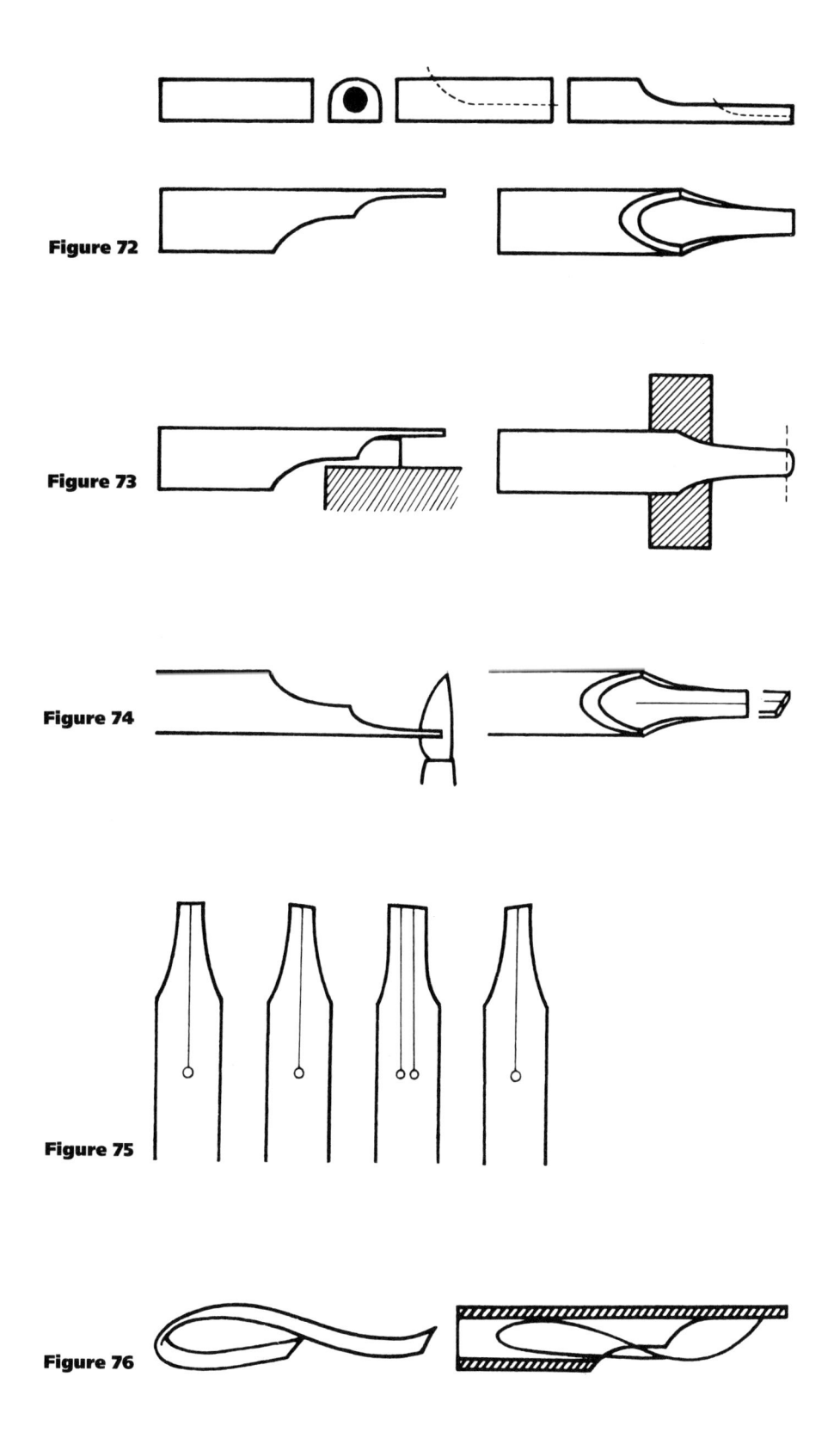

Figure 72

Figure 73

Figure 74

Figure 75

Figure 76

Brushes

Brushes are used for making large letters, for lettering on vertical surfaces and on surfaces that have a painted ground. Brushes can be categorized according to the shape of their bristles into round, flat, and pointed ones (Figure 77). Variations of all these forms are available.

Use round brushes for letters with equal stroke widths. The best brushes are made of red sable hair. A brush with a less than perfect shape can sometimes be saved by dipping the hair belly into water-soluble glue and rubbing it on a polishing stone until the desired form is achieved (Figure 78).

Mix the paint in a specially designated container, never in the original one. Do not use your lettering brush to stir the paint; an old brush will do the job. For large amounts of paint use a stick. There should be no lumps and the paint should flow freely. Dip the brush to the ferrule and turn it at the rim of the container to remove all excess paint except for a small drop at the hair point, which will form a perfectly round dot where the brush meets the paper. The beginning and the end of the stroke can be rounded separately with a round or flat brush; however, this technique is not compatible with most writing styles that are executed with a round brush.

Use flat brushes for letters with varied stroke thicknesses. Red sable provides the best spring. When the brush is damp, all the hairs should be in perfect alignment; any protruding bristles can be carefully trimmed into shape with a pair of scissors. The paint should be thin but not transparent. Dip the brush almost to the ferrule and wipe off excess paint on both sides at the edge of the mixing bowl. Hold the brush as if it were a pen with a right-angled nib. Since upstrokes tend to disturb the shape of the

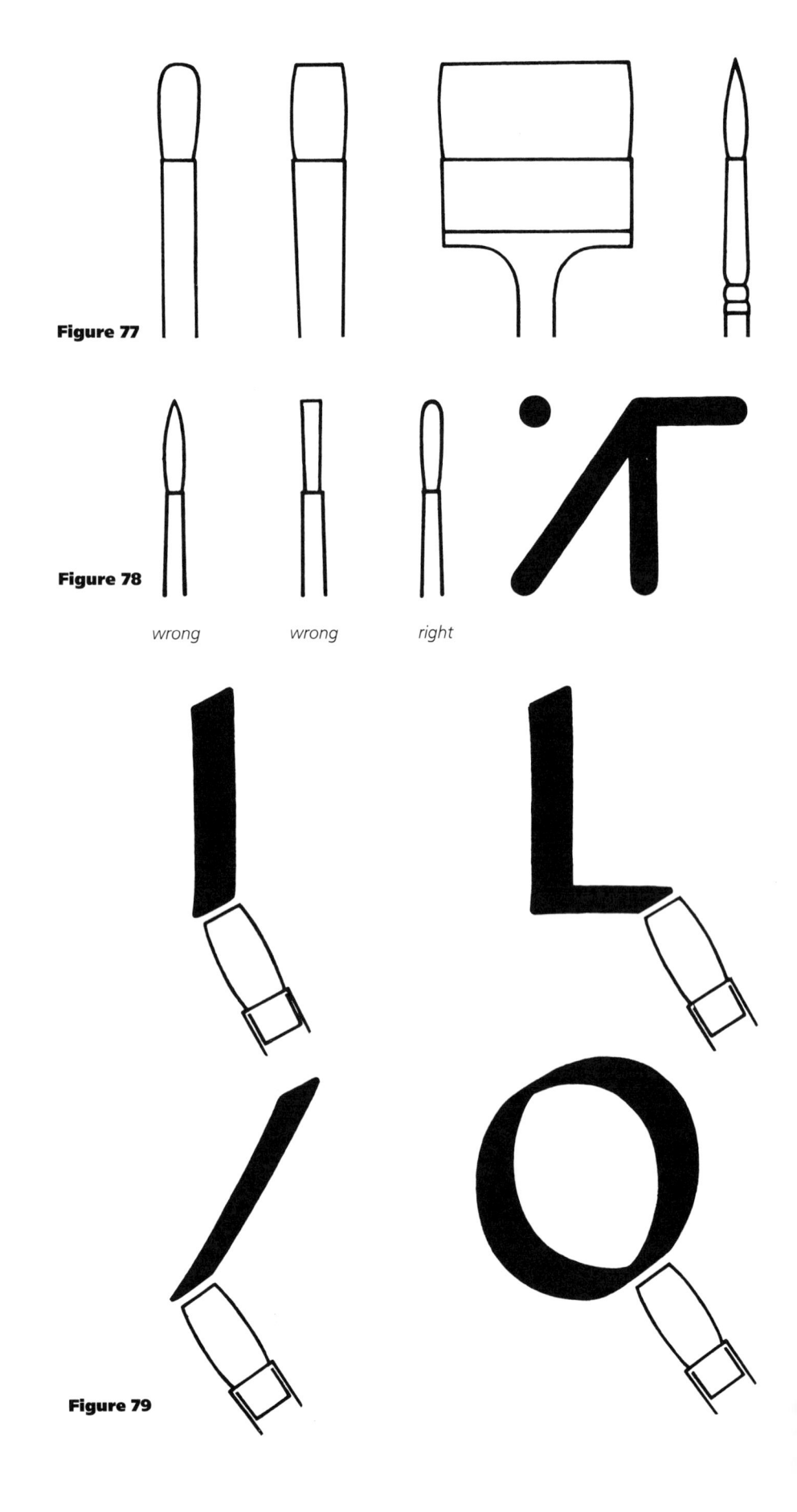

Figure 77

Figure 78

wrong   wrong   right

Figure 79

Figure 80

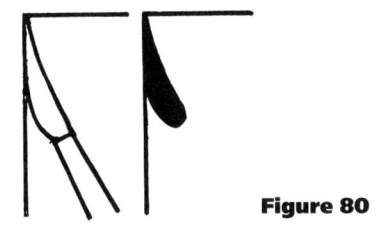

brush quickly, it is advisable to construct letters in as many downstrokes as possible. Even letters of equal stroke width can easily be executed with a flat brush, which can be turned to follow the shape of the letter (see Figures 93 and 115). Hog-bristle brushes come in different shapes and are best used for large letters. Writing long straight lines with a brush is easier with the help of a mahlstick (Figures 90, 91, 92).

Pointed brushes are most often used for drawing. The damp brush should have a perfect hair point; any hair that protrudes beyond it can be trimmed off with a razor blade. Dip the brush into the paint lightly and remove the excess on a glass or porcelain plate in a round motion. The stroke of a pointed brush starts in a hairline and swells with increasing pressure (Figure 80). When necessary, turn the paper rather than twisting your hand to complete the outline of a letter.

Take good care of your brushes. Wet them in water before each use and squeeze the water out again. This will prevent paint from settling inside the ferrule, which could later destroy the hair. Never let the paint dry on the brush; later attempts at removing paint can turn the hair brittle. If you use water-soluble paint, wash your brush in soapy water after each use. If you use oil or emulsion-based paint, clean the brush in turpentine first and rinse with soapy water. To preserve the shape of your brushes perfectly during storage, dip them into water-soluble glue and fashion a cardboard holder with elastic loops for each brush.

Other Writing Tools

An experienced letterer can use all kinds of implements, such as different-shaped pieces of wood, old hog-bristle brushes, felt-tip markers, chalk, and other materials to create many surprising effects.

Drawing Materials

Use a hard pencil (6H) and pointed sable brushes, preferably one for black and one for white. The best sizes are narrow, such as a number 2 or 3. Other useful items are tracing paper, a razor blade, a ruler, a metal square, pushpins, compasses with ink attachments, a magnifying and a reducing glass, and fine sandpaper mounted on a wood block to sharpen your pencils.

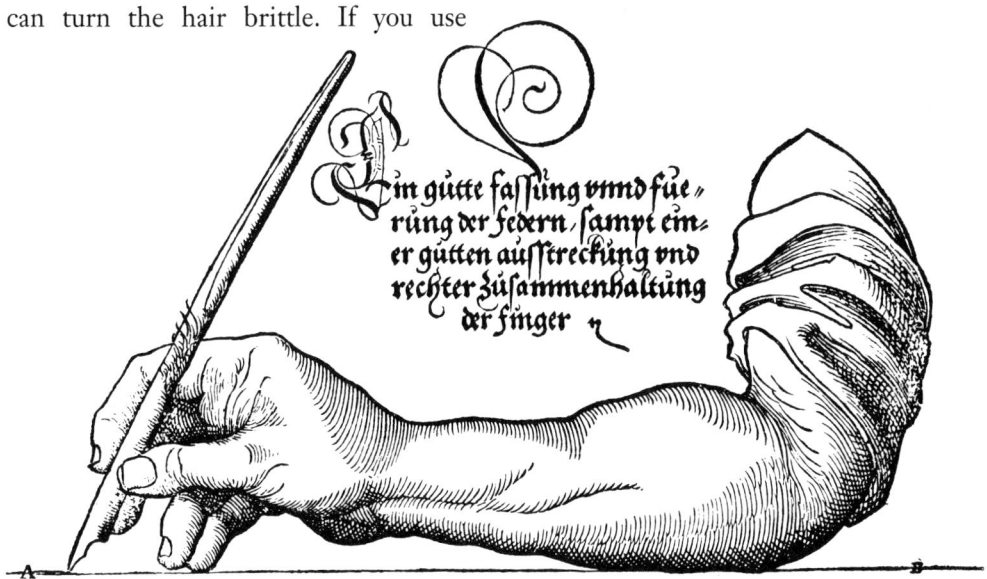

81 *The right way to hold a pen. From the facsimile edition of* Schreibbüchlein *(Little writing book), by Renaissance master Wolfgang Fugger. Leipzig, 1958.*

82 *Condensed sans serif alphabet by Heinz Schumann.*

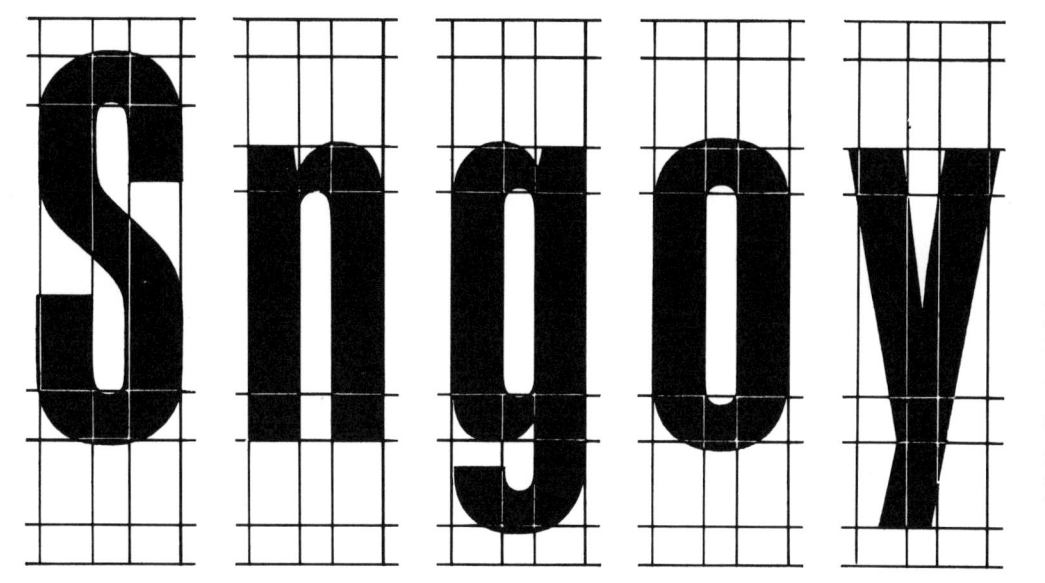

Figure 83

Figure 84

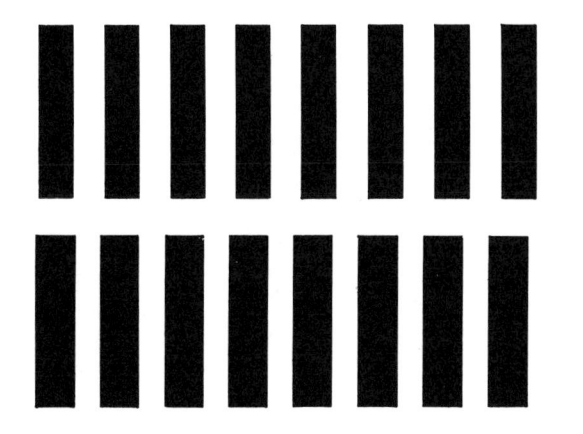

Figure 85

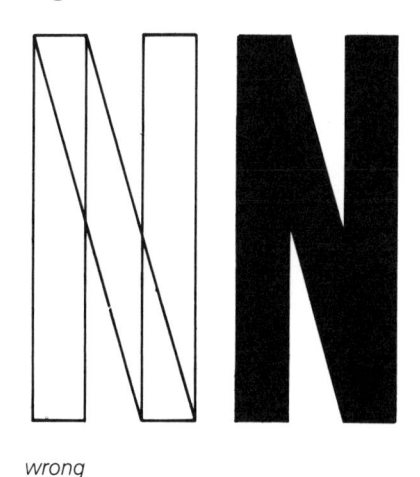

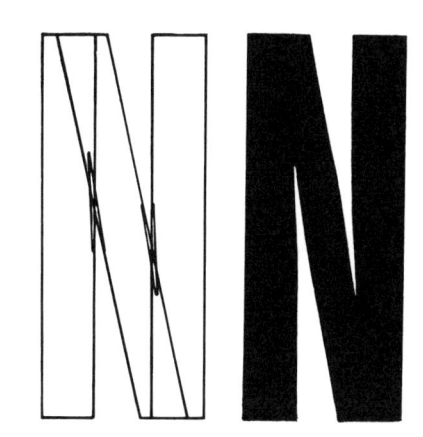

wrong

right

Figure 86

EXERCISES

Condensed Sans Serif

Condensed sans serif is frequently used in design projects, because its proportions can be varied freely to allow many decorative effects. Its basic form is simple and can be adapted to different techniques. It is easy to master and therefore a good choice as first project. The basic scheme of its construction is the transformation of round forms into straight lines with rounded corners (Figure 83).

Drawing and Cutting

Our first project is to draw the letters, cut them out, and paste them back onto paper. Chose white, black, or colored paper, which should be heavy enough to withstand handling but light enough to produce a clean edge when it is cut. Black paper should have a light back side that will allow you to see pencil lines. The outline of the letters can be drawn in one of two ways. A mirror image of the letter can be drawn directly onto the back of the paper, or the original drawing can be made on tracing paper and transferred to the back of the paper in mirror image or to the front of the paper as a direct copy. If you choose the latter method, check to see if erasing will leave marks on the surface of the paper, a danger especially when light-colored papers are used. Useful measurements for these exercises are as follows:

height of uppercase letters: 2½ inches (6.5 centimeters)

x-height of lowercase letters: 1¾ inch (4.5 centimeters)

Since the condensed sans serif ascenders are short, its uppercase letters, also known as majuscules or versals, do not have to be shorter than the ascenders, as is common in classical roman letters. The stroke width of the lowercase letters, or minuscules, should be one-eleventh to one-twelfth narrower than that of the uppercase (see Figure 34).

We will begin with the uppercase O and H and with the minuscules m and o. On these letters the width of the vertical and horizontal strokes, the width of the internal space, and its relation to the height of the letter will be established. A tall and narrow letterform usually appears more elegant than a heavyset and wider one. The space between verticals should either be slightly wider or slightly narrower than the stroke width (Figure 84). If this rule is ignored, the image will flicker in front of the viewer's eyes. Remember the discussion on the relationship between black and white areas of equal size in Chapter 1. If the lettering is white on black or any other colored ground, adjust your measurements accordingly.

For the first exercise choose a space between letters that is narrower than the stroke width; this arrangement makes proportions clearer and easier for beginners to handle.

Draw the horizontal lines somewhat thinner than the verticals. In round parts keep the upper and lower parts narrower than the side parts (see Figures 29, 30, and 32).

It is likely that you will have to make several attempts at the shapes of H, O, m, and o before you reach a satisfactory arrangement of stroke widths and space between letters and before you can proceed to other letters.

Naturally you have to adapt the construction scheme for the letters A, V, M, N, v, and w with respect to their widths. Letters that are made up of several strokes appear darker than letters with fewer strokes. A letter B will require a thinner vertical line than a J, I, or L. The inner diagonals of N, M, and W have to be considerably thinner than the outer lines. The area along which a vertical and a diagonal line meet should be as wide as possible. This creates space for the interior angles and avoids dark spots in the design. Figure 86 illustrates this principle. For the same reason angles should be drawn with exaggerated points (Figure 85 and see Figure 35, page 22). It is a characteristic of the condensed sans serif that there are many transitions between straight and rounded forms. The effect of this typeface depends to a great deal on the successful management of these transition points (see Figures 36 and 37).

The point at which the curved part meets the main stem is of special importance in letters like n, b, and h (Figure 86). Keep the angle deep enough to be visible in a photographic reduction of the letter to a size of $3/16$ inch (5 millimeters). Use a reducing glass.

A number of optical illusions and other problems may arise which cannot be understood in depth until the student has progressed further. Figure 82 shows a well-developed alphabet of condensed sans serif. The designer here deviated in several instances from the original scheme, but for the beginner it is useful to identify the already discussed points and to pay close attention to careful drawing and cutting.

When cut letters rest loosely on a surface, they will produce shadows along some of their edges. Cover them with a sheet of glass for more precise control.

Work on the letters in the following sequence: H, M, E, O, A, N, U, B, R, S, h, m, n, r, u, o, b, e, a, s, g. Manipulate these key figures until they satisfy your demands with respect to stroke widths, relationship of shapes, and width of the letters. Frequently it is necessary to draw several versions of a letter and decide from the interaction with other letters which version is the best one. Arrange your cut letters into words before you

On page 41:

90 *a and b Holding a mahlstick: different finger positions.*

91 *Drawing a line with a mahlstick on a horizontal surface.*

92 *Drawing a line with a mahlstick on a wall.*

93 *How to hold a flat-tipped brush.*

Figure 87

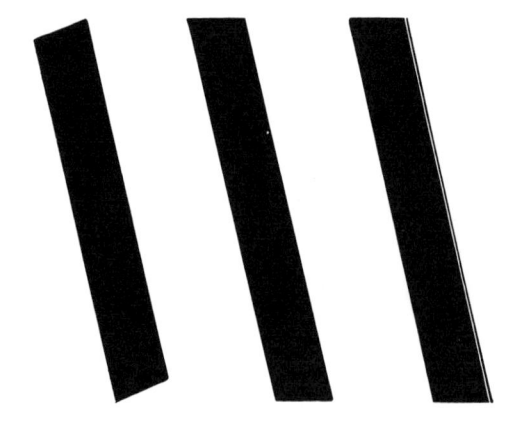

Figure 88   **Figure 89**

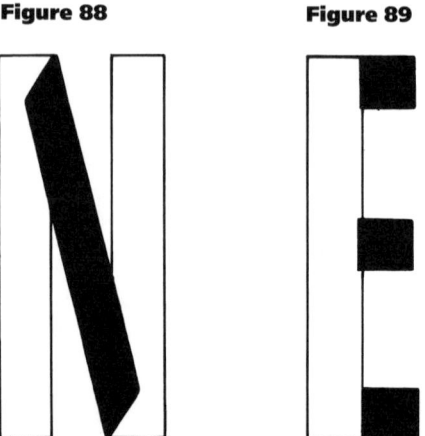

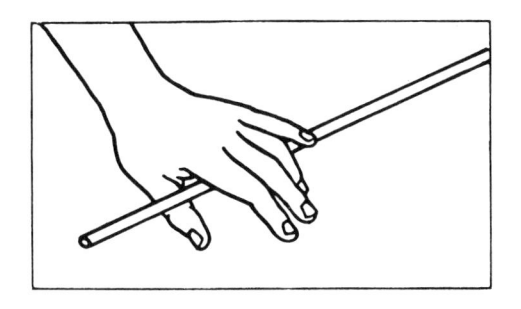

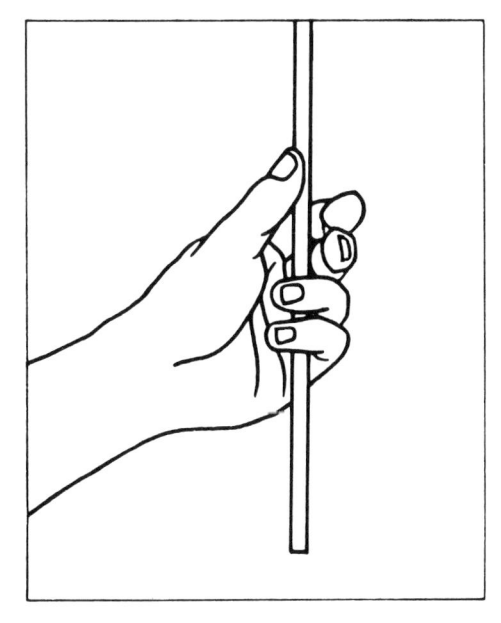

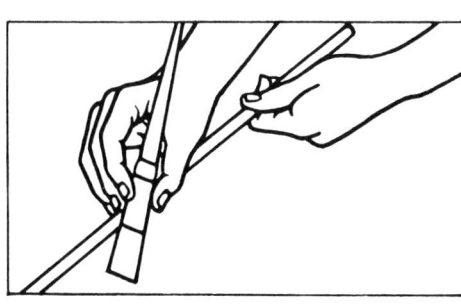

Figure 90a and b, Figure 91

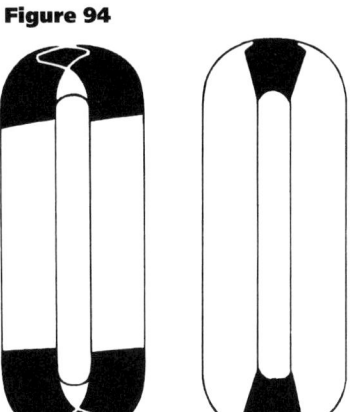

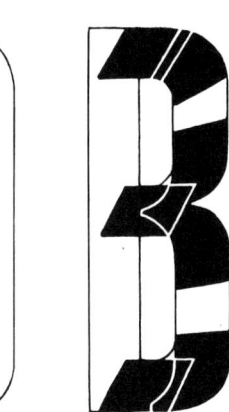

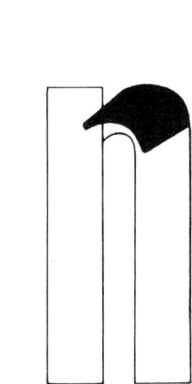

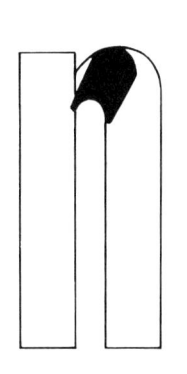

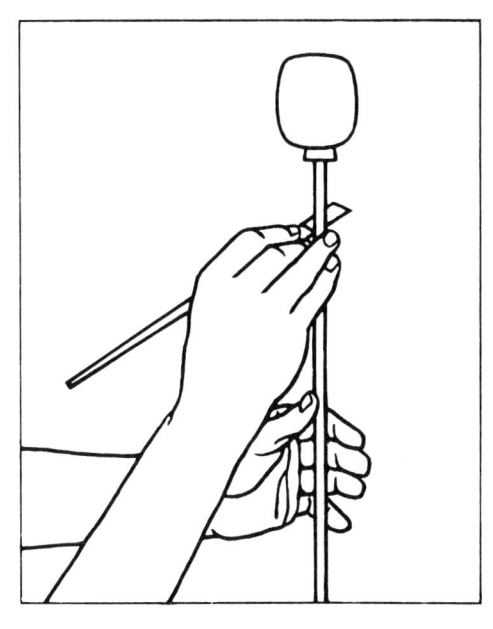

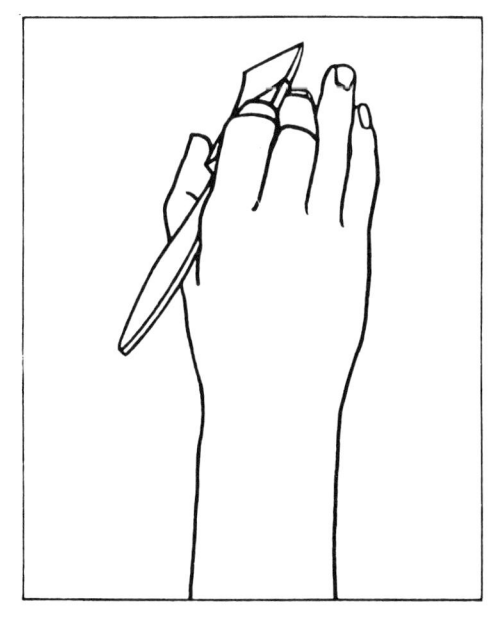

Figure 92, Figure 93

Figure 94

Figure 95

glue them in place, and check their size, proportion, relationship to each other, gray values, and their overall beauty. No single letter should stand out as a strange element.

Use gum arabic to glue the letters down. It will not cause the paper to expand and later contract in wrinkles, because it will not be absorbed. Apply the adhesive only on strategic spots and you will be able to make changes if necessary. Visible spots of gum can be removed with an eraser.

Lettering with a Flat Brush

A flat brush is an appropriate instrument for lettering condensed sans serif. Use a mahlstick to draw straight lines. The technique for using it on a vertical or horizontal surface varies, and may pose problems for the beginner, but it is advisable to attempt the method shown in Figures 91–93, since this will be most useful in the long run. A horizontal surface should be brought to an angle of at least 30 degrees. Wrapping paper is the most convenient material.

Start by practicing strokes in the size and width of an I. Hold the edge of the brush horizontally when you draw a vertical line; hold it almost vertically or slightly slanted when you draw a horizontal line. Draw the beginning and end part of the strokes securely and with determination. Subsequent attempts to patch a timidly drawn area will always be visible.

Use the mahlstick at a slant for diagonals and finish the corners at the end. The edge of the brush can also be held horizontally from the beginning of the stroke to the end. Some pressure or a second stroke, slightly overlapping the first, may be necessary to achieve the desired width (Figure 87).

The letters M, W, and N present special situations because their inner diagonals are narrower. Here you have to

41

pull the edge of the brush at a steeper angle (Figure 88).

When you make small sans serif letters, it is more convenient to draw all horizontal bars as vertical lines (Figure 89).

For the letter O use a mahlstick for the vertical sections and form the round parts using three strokes each. Hold the brush at a slant and start with the almost horizontal sections of the side parts. The letters S, G, C, B, e, s, and c are constructed in a similar manner (Figure 94).

The curve of letters like n, m, b, and h consists of two strokes, both executed with a slanted brush (Figure 95).

Use a brush with stiff bristles for letters with stroke widths of more than ¾ inch (2 centimeters). With some practice it is possible to draw the round sections freehand in sections of a quarter circle each (see Figure 102).

Even-stroke Sans Serif Roman

Uppercase Letters

Drawing versus Constructing. The square, triangle, and circle supply the basic shapes for the Latin alphabet and the roman letters of even stroke width are a prime example. Since there is no rhythmical change between thick and thin lines, no serifs or other decorative elements, even beginners have no difficulty in recognizing the proportions of the letters. Figure 101 serves as nothing more than a starting point. Take a soft but well-sharpened pencil and practice drawing the letters on graph paper until the proportions seem right (Figure 99). Now let us take a closer look. The construction process itself is not the aim of the exercise, but merely a vehicle. The shapes cannot be forced into a mold; they should develop naturally, and a system of construction rules is helpful in the process.

Attempts at the geometric construction of letters with ruler, compasses, and measurement units have been made since the Renaissance. We find well-known examples of such letters in Pacioli's *Divina Proportione* (1509), in Dürer's *Unterweisung der Messung* (1525), in the *Romain du Roi* from the end of the seventeenth century (Figure 97), and in the sans serif of the Bauhaus artist Herbert Bayer (Figure 98). None of these attempts produced an aesthetically satisfying result. Form develops out of movement, which can only be stifled by excessive construction. The act of writing is the basis of every letter; organic movement is the essence of its form. Complex lowercase letters in particular defy strict construction rules, and the only reliable guides are examples, experience, and a trained eye.

Draw your letters about 2¼ inches (6 centimeters) high. Choose a stroke width of at least one tenth of the height, but not wider than one sixth of it. Wider strokes would destroy the classic proportions of a sans serif roman letter. Extremely wide strokes would require a change in the width of the letter in relation to its height (Figure 100). The letters E and B will thus appear as wide as H and Z. A combination of both would be stylistically unacceptable.

Draw straight lines with a ruling pen and fill them in with a brush. If you construct curves with compasses, do not forget to correct the shape (Figure 102), and take into consideration the remarks on optical illusions in Chapter 1. You will achieve the best results with curved letter parts if you draw them freehand with a brush.

Figure 96

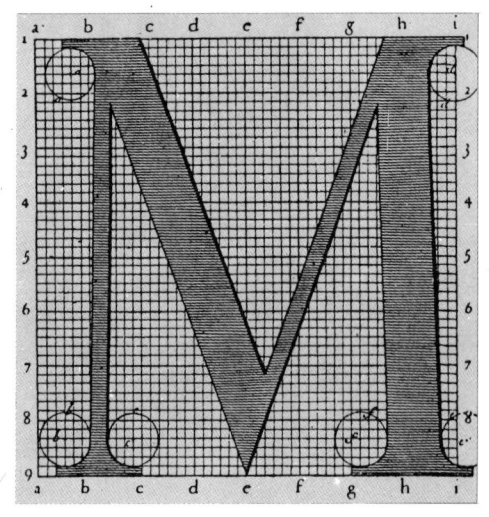

Figure 97

Figure 98

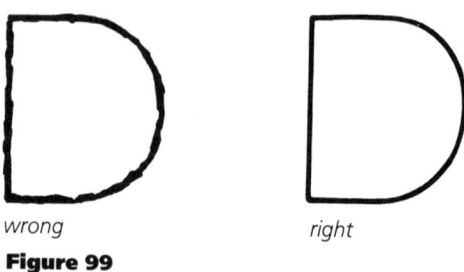

wrong right

Figure 99
Figure 100

On page 42:

96 *Construction of a Renaissance capital letter. From* Schreibbüchlein *by Wolfgang Fugger.*

97 *Drawing from* Romain du Roi, *1692.*

98 *Experiment in lettering by Herbert Bayer, about 1926.*

99 *Drawing a letter.*

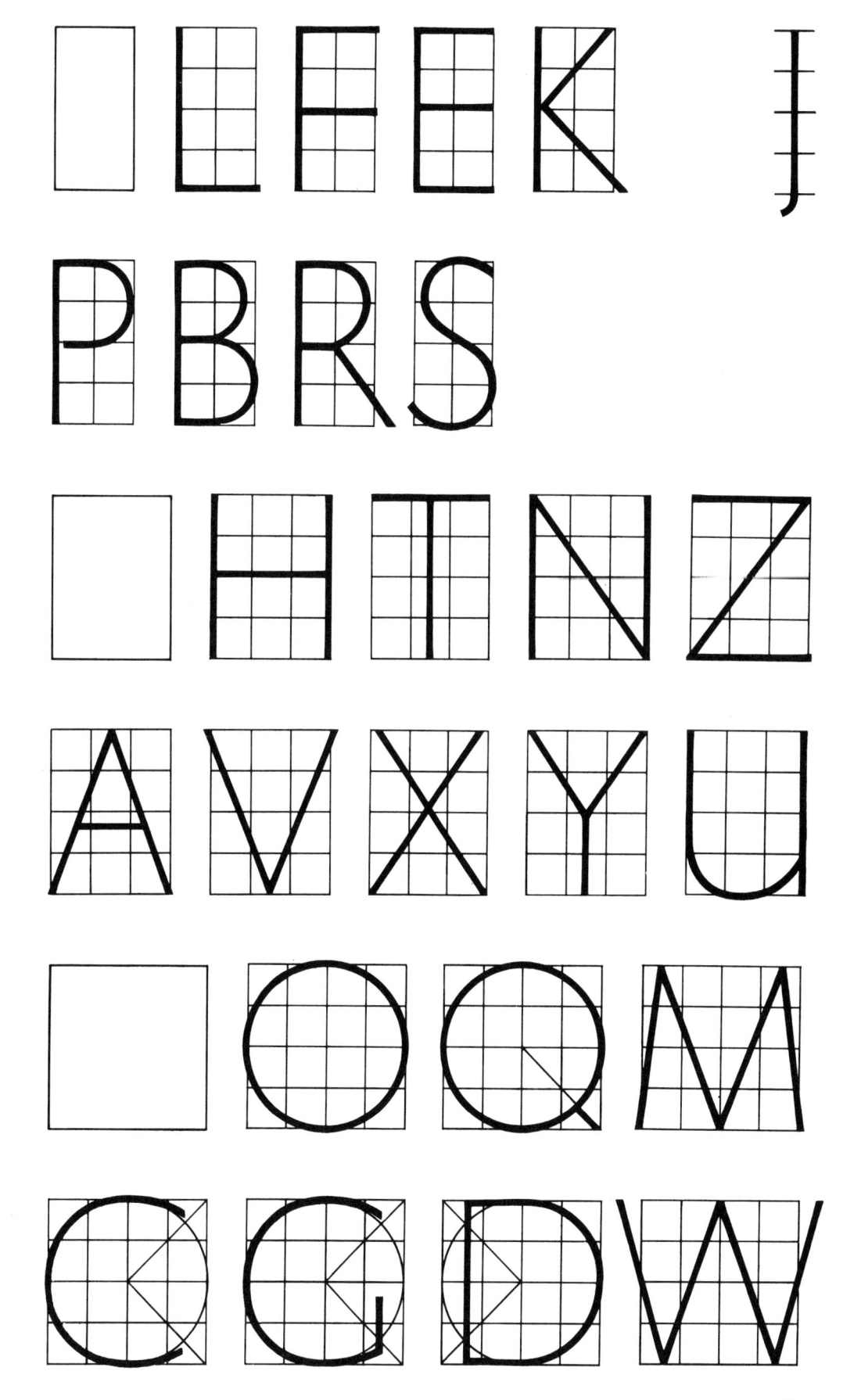

Figure 101 on this page

Figure 102 on pages 44, 45, and 46

43

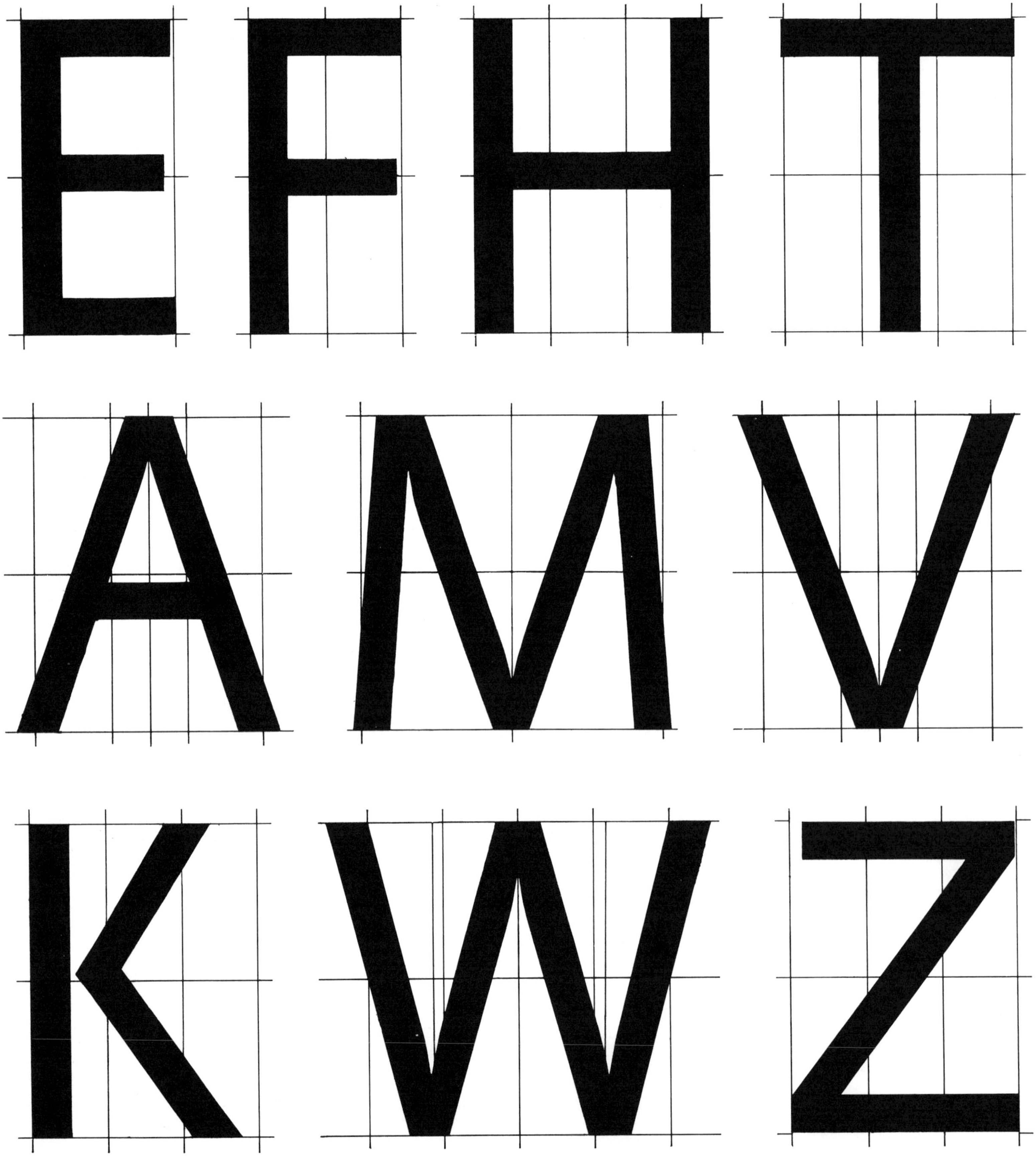

45

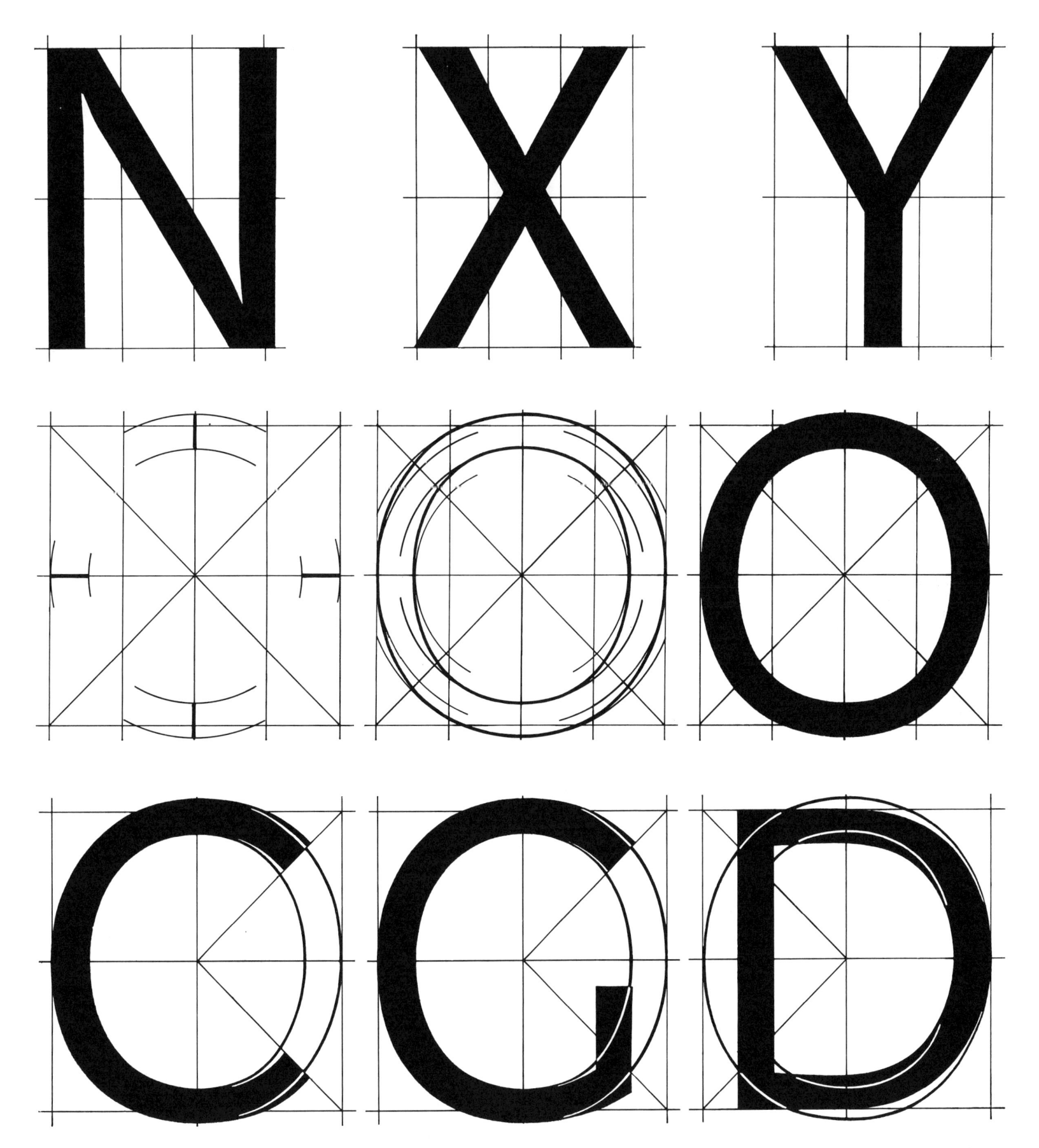

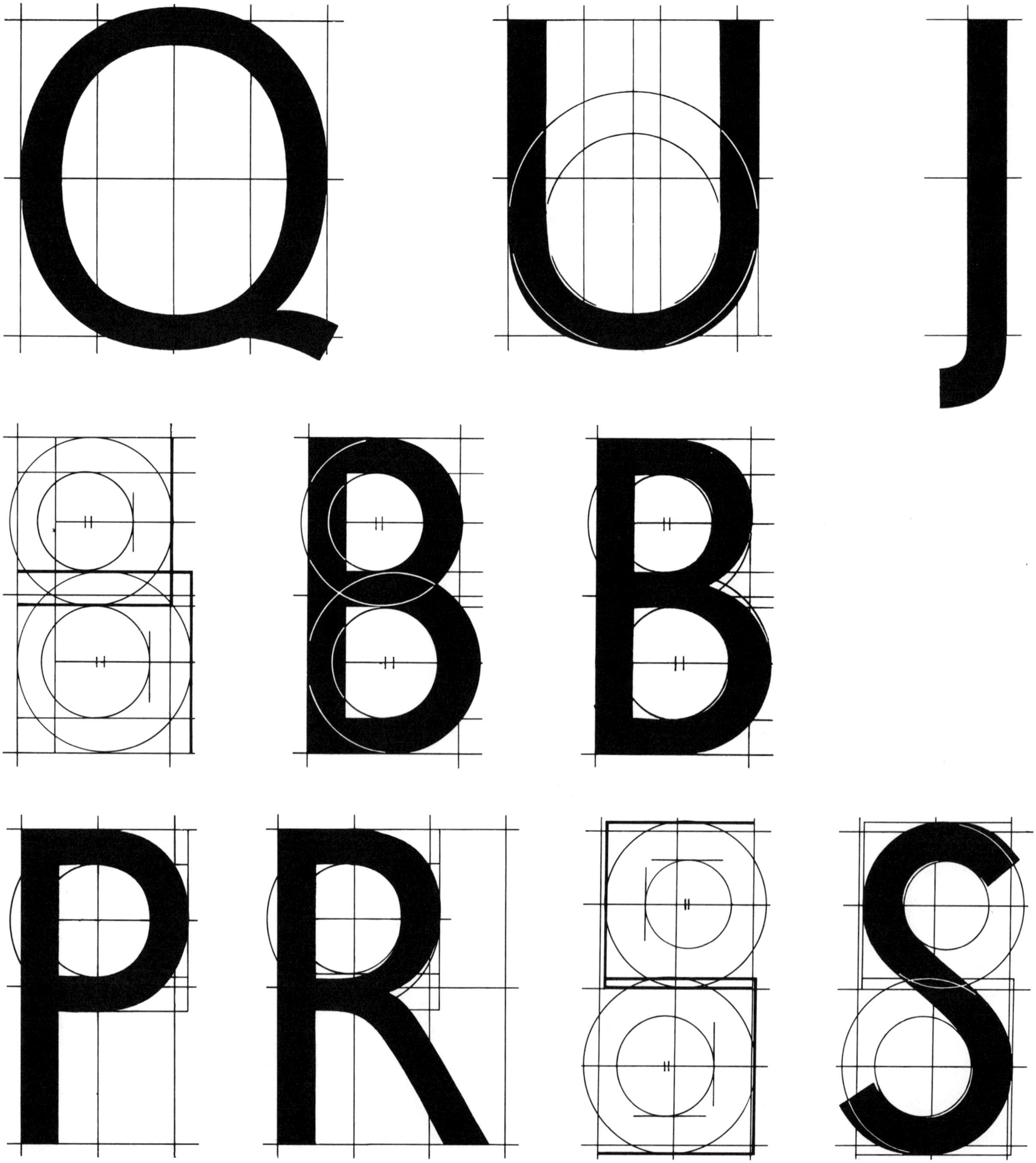

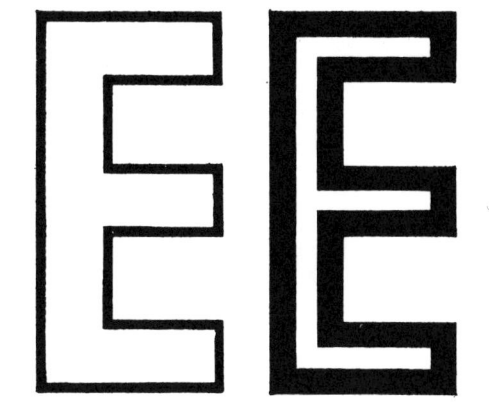

103 *Even-stroke roman.*
104 *Decorative variations.*

Figure 105

Figure 106

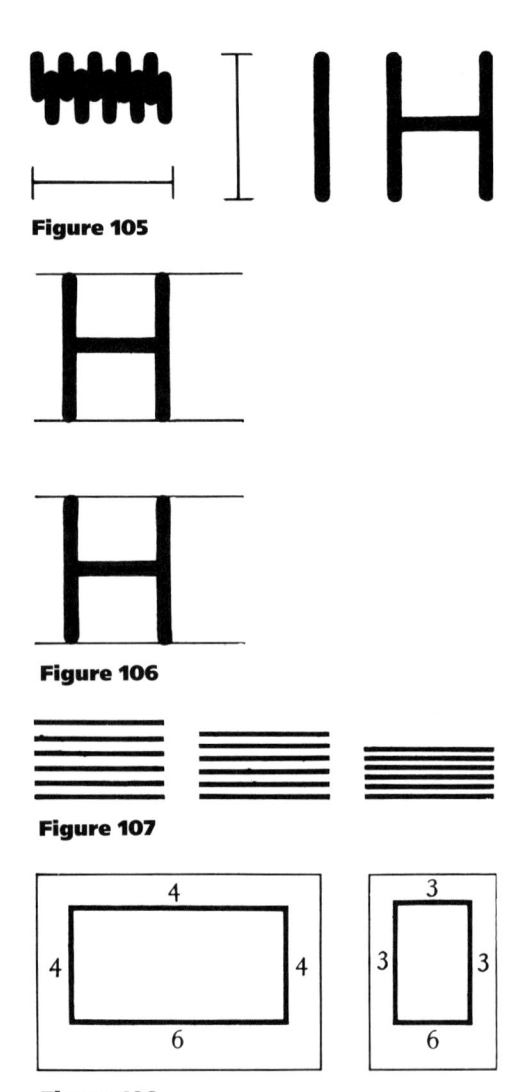

Figure 107

Figure 108

Figure 109

Lettering with a Round-ended Nib. Before you start work, it is important to find the right ratio between stroke width and letter height. The measurements given on nibs are not necessarily accurate and should be checked (Figure 105).

A ratio of one to ten seems best for first exercises. If you use a pen that makes lines 1/16 inch (2 millimeters) wide, choose a letter height of 3/4 inch (2 centimeters). Next you have to determine the amount of space between the lines. Not all the hints given in Chapter 1 will be relevant to the beginner, but it is advisable to choose a distance of slightly more than half the letter height (Figure 106) at first. Subsequently one of two methods can be used: letter groups of three or four lines each with different line intervals and choose the spacing that looks best, or cut out the lines and rearrange them until the right spacing is achieved.

Determine your margins on an exercise sheet and draw guidelines on it (see The Work Space, page 33, and the section on Documents and Addresses, page 175). Consider the relationship of the text block to the margins (Figure 108). It is not necessary to aim for justification—alignment of the lines at the right margin—and the lines certainly should never be squeezed into a block of predetermined size. An unaligned and somewhat restless right margin is usually more pleasing than letters and spaces that have been stretched or compressed to fit exactly into the full measure of the line. Both the left and the right margin, however, should carry the same weight optically.

Use the basic form elements of the capitals as components for ornamental exercises. This will familiarize you with your pen, and you can experiment with distances between the elements. Let round and pointed shapes protrude slightly from the guidelines (see Figure 22). Combine the rows to design areas, and aim for a dense texture on the page as well as for even gray values over the entire area (Figure 109).

Figure 111 shows the proper direction of each stroke. To make the learning process as simple as possible, practice letters with similar elements as a group. Start with those that consist only of vertical and horizontal lines: I, E, F, L, T, and H. Remember the remarks about the optical middle of objects in Chapter 1, and place the crossbars accordingly. The only exception to this rule is the letter F: its short middle arm can remain in the geometric middle of the vertical.

Next comes the group of letters that contain diagonals: A, V, N, Z, M, X, W, Y, and K. Make little dots at the desired endpoints of the diagonal lines and connect them later, but try to write the letters as unencumbered by such construction help as possible. Difficulties usually arise with M and W. The two outer lines of M are neither exactly vertical, nor are they as slanted as the two outer lines of W. The M is not an inverted W! In the letter W the left-and right-slanting diagonals should be parallel to each other. The space inside the Y is frequently too small; make it slightly larger than the space enclosed by the upper portion of X.

The third group consists of letters that are roughly a circle in form: O, Q, C, G, and D. None of these letters forms a perfect circle; the upper halves appear larger (Figure 112).

The last group is made up of B, P, R, J, U, and S. The crossbar of the B lies in the optical middle; the interior spaces of R and P should be larger than those of B, which brings the crossbars of R and P into the geometric middle.

Keep comparing your letters to models or established proportions. Faulty designs usually attract attention through

ABCDEFGHIJ
KLMNOPQ
RSTUVWXYZ
1234567890

110 *Sans serif even-stroke roman.*

111 *Stroke directions.*

ABCDEFGHIJKLMN
OPQRSTUVWXYZ
1234567890

OGD H H H

Figure 112 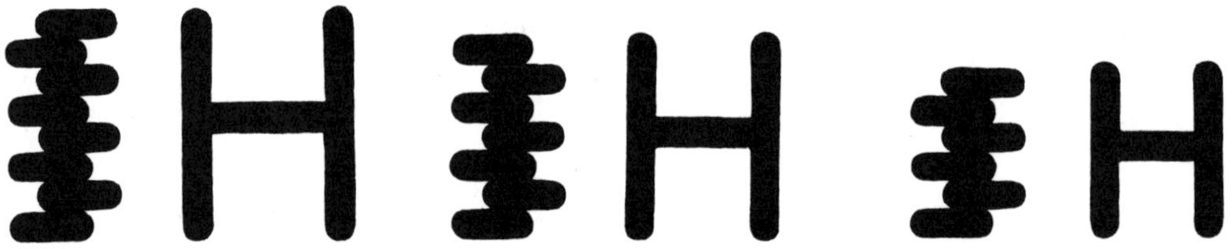 **Figure 113**

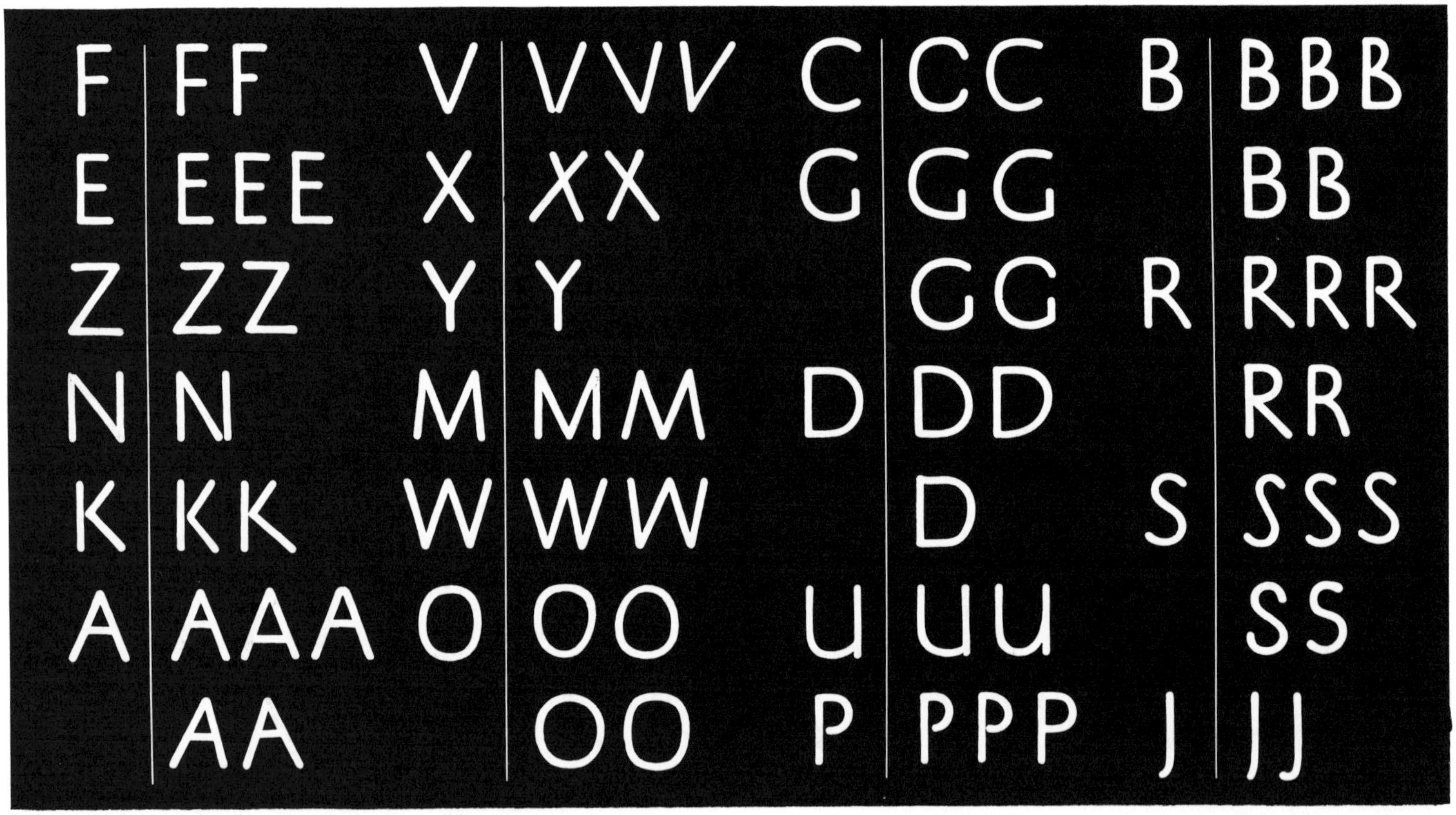

F	FF	V	VVV	C	CC	B	BBB
E	EEE	X	XX	G	GG	B	BB
Z	ZZ	Y	Y	G	GG	R	RRR
N	N	M	MM	D	DD		RR
K	KK	W	WW		D	S	SSS
A	AAAA	O	OO	U	UU		SS
	AA		OO	P	PPP	J	JJ

Figure 114

poorly proportioned interior spaces. Figure 114 shows some of the most common mistakes.

Check the way you hold your pen. If you make errors in writing, do not try to correct them; let them stand and measure your improvement against them. Some letters are so difficult to write that you will need considerably more time to master them than others, but resist the temptation to fill lines and pages with an endless repetition of the same form. Do not add alphabet to alphabet in series or in vertical alignment of the same letters. Write words, and pay attention to groupings of letters and the spaces between them (see Spacing, page 23). When you are comfortable with one group, proceed to the next, then mix the letters; rearrange them into other words and later into running text. Finally, practice the letters in various stroke widths (1:8,1:7,1:6). Only the light and medium weights will look good; if the strokes are too bold for the size, you have no leeway to counteract optical illusions.

Lettering with a Brush and Other Tools. If you practice lettering not merely as a hobby but with more serious intentions, you should become familiar with the use of a brush. Letter on wrapping paper, on paper samples, or on any other cheap material. Rough surfaces resist the pull of the brush and create interesting effects. Letter on a variety of papers and on surfaces that are primed with paint. As before, it is best to work on a slanted table: the working surface should be propped up at an angle of at least 30 degrees.

Strokes up to ¼ inch (6 millimeters) wide can be made without the aid of a mahlstick. The section on Brushes, page 36, describes the right way to handle a brush. It is very important that you wipe the brush on all sides after you dip it. Do not push down while you write, and keep the ends of all crossbars nice and round and of equal thickness. The rounded ends can be corrected with a flat brush if absolutely necessary.

Stroke widths of more than ¾ inch (2 centimeters) can be executed with a pointed brush, but the result is often aesthetically disappointing. For large letters, flat brushes usually yield better results.

Liner or script brushes with their longer hair belly can be used for large letters, but their use requires a great deal of practice, since it is easy to twist them out of shape at round sections of the letters. Use bristle brushes for letters that approach ¾ inch (2 centimeters) in width. Grip the tool near the ferrule and guide it with a movement from your shoulder and your elbow. Neither your lower arm nor your wrist should rest on the paper: only your pinky can be braced against the surface. Use a mahlstick for making straight lines, if you wish (Figures 90–92). Hold the edge of the brush horizontally for vertical lines and vertically for horizontal lines; for cursives, turn the brush (Figure 115).

Practice using a flat brush to make a condensed sans serif (see Condensed Sans Serif, page 39). Here interior spaces are small and technical difficulties manageable.

Even-stroke roman letters acquire a new and interesting aspect when they are made with different instruments. Figure 116 shows letters made with a watercolor brush; the letters in Figure 117 were drawn with a felt-tip marker, and those in Figure 118 with a wooden stick. Letters cut in wood have a distinct character (see Figure 423, on page 205). Yet another effect is achieved with positive images cut in linoleum, attached to wood blocks, and used as stamps. And letters can be cut or ripped from the background material, as shown in Figures 119 and 120. These suggestions are intended only to inspire the beginner to his or her own attempts.

Figure 115

116 *Watercolor brush.*

117 *Felt-tip pen.*

118 *Wooden stick.*

119 *Cut from paper without a pattern.*

120 *Torn from paper.*

51

abbcdefghij klmnopqr stuvwxyz

121 *Lowercase letters of a sans serif even-stroke roman.*

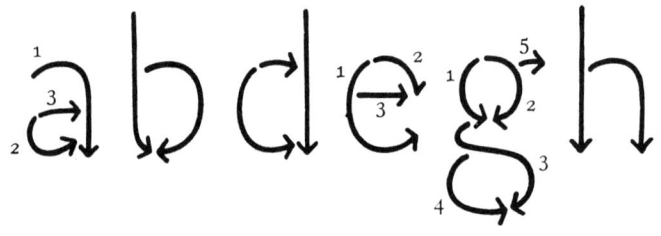

122 *Ductus, or stroke sequence and directions.*

aaabbdddptfjk mnnuusseee

123 *Poor letterforms.*

Figure 124

Figure 125

Lowercase Letters

Minuscules were developed from majuscules over a long period of time. Several form changes took place which can only be appreciated when the historic context becomes clear. We are turning our attention to lowercase letters at this point because they make up the vast body of printed and written material in our time. Long texts made up entirely of capital letters are difficult to read.

Figure 124 shows the proportions of stroke width to height: the x-height is about six and a half times the stroke width. Capitals of the same type should have a height of nine stroke widths. The ascenders are somewhat taller than the capital letters, which would otherwise appear too big and cause optical holes in the page design. Mark the waistline, which defines the upper end of the x-height, with a pencil line. Ascender, descender, and cap heights are best estimated.

Lowercase letters are grouped differently from capitals. Practice m, n, h, i, u, and o first, and tune their widths to each other. Write your m with smaller interior spaces than n. Do not represent o with a circle, and practice related forms along with it: b, d, e, g, p, and q. Keep writing m, n, h, and u, and add the newer letters to your repertoire. The letters k, v, w, x, y, z, s, c, and l should not pose any new challenges, since they correspond closely to the capital forms. The last group consists of a, t, and j. Make the t with a flat bottom arch; treat j in a similar way, and place the crossbar of the t somewhat below your auxiliary pencil line. Common mistakes are shown in figure 123. With some practice the lowercase even-stroke roman letters can be made easily with a flat brush. Special attention is required at the spots where arched forms meet straight lines. Avoid blobs of ink at these points by thinning the lines before they meet. To

right	nib edge too flat	nib edge too steep

Figure 126

Figure 127

right	wrong	wrong	right	wrong	wrong

Figure 128

Figure 129

right	wrong	right	wrong	wrong

Figure 130

wrong	right

Figure 131a

wrong	right	wrong	right

Figure 131b

achieve this effect, simply lift the brush off the paper slightly (Figure 125).

Sans Serif Roman with Different Stroke Widths

Uppercase Letters

Capitals made with different stroke widths are subject to the same considerations of proportion as the even-stroke letters. The angle of the pen is of utmost importance, however: it determines the character of the letter (Figure 126).

After you have decided on the stroke width and letter height, proceed in the same manner as for the sans serif (Figure 127), and practice the basic strokes (Figure 129). Try to give each stroke a definite beginning and end. Touch the pen nib to the paper with some pressure and lift it slightly to form the stroke, but make sure that it always remains in contact with the paper. Increase the pressure slightly again just before you lift the pen off the paper (Figure 128). All horizontal lines are formed in this way. Too much pressure will result in black spots at the end of the stroke and should, of course, be avoided. Corners look better if the horizontal lines slightly overlap the verticals to which they are attached (Figure 130). Where two diagonals meet, the wider stroke can also overlap the other one.

For round shapes hold your pen as you would for a horizontal stroke, but make the widest part of the stroke in the middle of the movement and end the stroke in a point. A diagonal axis will result.

Practice separate shapes by combining them in rows of ornaments, and add further interest to the exercise by using red or blue ink for some rows. For a beginner, this new challenge of using and distributing different hues is of great value.

53

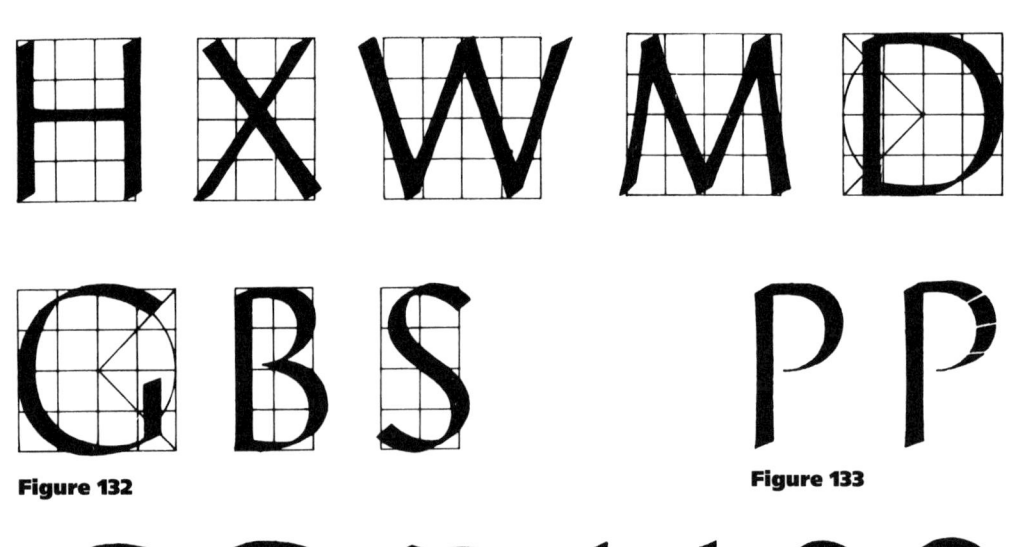

Figure 132   **Figure 133**

134 *Incorrect forms.*

135 *Stroke direction for serifs.*

Practice the letters in the same groups as you did for even-stroke roman. The letters M, N, and Z often seem difficult. If you hold the pen in normal position for vertical lines and for the diagonal of the letter Z, then the letters will appear too heavy, but the Z will appear too light. Therefore you should turn the pen to make the first vertical line of the M and the two verticals of the N as wide as the horizontal lines of the other letters; make the diagonal of the Z full width (Figure 131).

The two curved strokes that make up the O should connect without any visible gap. Long thin sections are equally undesirable. Any mistakes are most clearly visible in the interior space. Keep the arches of capital C and G relatively flat.

The letters R and B require much practice. Start the half circles with horizontal lines and lift the right side of the nib off the paper near the end of the stroke to avoid cramped forms in an area that is small in relation to the stroke width. Another way to make small curves is to gradually adjust the right angle of the nib until it reaches a horizontal position (Figure 133, right). When the ends of small strokes in R, B, and K reach too far into the vertical stems of the letters, a cluttered image results. Deformed interior spaces ruin the entire form. Some of the most common mistakes are shown in Figure 134.

Attractive variations of this alphabet can be created through the addition of serifs or serif-like forms at the beginning and end of the strokes (Figure 137 and 138). Entry and stem as well as stem and exit can be executed in one stroke. The entry is relatively flat, and the form that leads into the straight line is achieved by a slight increase of pressure (Figure 141). For beginners it is advisable to make the serifs that come at the end of

ABCDEFGHIJ
KLMNOPQ
RSTUVWXYZ
1234567890

136 *Sans serif roman capitals with varying stroke widths.*

137 *Variation of the alphabet in Figure 136.*

ABCDEFGHJ
KLMNPR
STUVWXYZ

ABCDEFGHIJ
KLMNOPQ
RSTUVWXYZ

138 *Another variation of sans serif roman with varied stroke widths.*

abcdefghijkl
mnopq
rstuvwxyz
1234567890

139 *Lowercase sans serif roman with varied stroke widths.*

nmomonhomunomon
mohemcmsmbntuelor

a stroke in two different movements.

A pleasing rhythm comes with practice. For variation choose different heights, stroke widths, and tools for your work, but remember that steel nibs are usually best suited to small letter sizes. Very impressive larger forms can be created with reed pens and brushes, the latter requiring disciplined practice because the width of the instrument changes as pressure on it varies.

Lowercase Letters and Numerals
As with capitals, the height of lowercase letters is determined by the stroke width. Use six stroke widths for the x-height, ten for letters with descenders or ascenders. Capitals of the same alphabet are written at a height of eight to nine stroke widths (Figure 140).

Practice the lowercase letters in the same sequence that was established for the sans serif. The key letters that determine form and width for all the others are o and n. The round forms, even though reminiscent of circles, show slightly more expansion in the upper half. If the pen is held in the right position, the axis will slant diagonally through the interior form. Start and finish verticals with serifs (Figure 141).

Make numerals the same size as uppercase letters and align them on the guideline or arrange them to create ascenders and descenders (as in Figure 139), which makes groups of numbers more legible.

Figure 143 shows letters with various errors. You will not be able to transfer directly to your work with lowercase letters with serifs all the information that you have gained from earlier practice. Changing stroke widths and serifs influences the optical appearance of the space between the letters, and new adjustments are necessary.

Figure 140

141 *Stroke direction for stems.*

Figure 142

143 *Incorrect forms.*

Italic

Lowercase Letters

To learn how to write an italic script is a much simpler task than most people imagine: it usually involves just a modification of the script taught in school. Figure 144 shows the handwriting taught in German schools, which resembles that taught in the United States and elsewhere. Historically, it developed from the humanist scripts of the Renaissance (Figure 145).

We start with lowercase letters, because they determine the character of this alphabet. In the beginning use a broad flat-edged pen, and try a flat or round brush later on. If the angle between the edge of the nib and the baseline is chosen correctly, a particularly rhythmic and dynamic image results from the change of thick and thin strokes.

The round forms of the italic are based not on a circle, as was the case with roman letters, but on a slanted oval. This produces slender letters of very similar widths. While the basic construction of roman letters consists of connecting round and straight elements that remain isolated from each other, italic script consists of a flowing up and down movement. Avoid isolated strokes wherever possible: your lettering speed will increase.

Practice the basic elements first.

1. Make downstrokes at an angle of 75 to 80 degrees, upstrokes somewhat steeper (Figure 146). Set a distance between the basic strokes that is neither too narrow nor too wide.

2. Make upstrokes in one movement, as shown in Figure 147. The upstroke can be smoothly connected to the downstroke (1) or overlap it slightly (2). The overlap should cover approximately half the x-height. Variations can be made ac-

144 *German school script, 1968.*
145 *Italian Renaissance italic by Palatino.*

cording to personal taste. Make arches not too roundly, but certainly never with points or broken lines. Figure 148 shows wrong forms.

Make upstrokes short, and match them to the size of the nib (Figure 149).

Make sure straight lines with rounded ends do not deteriorate into S-curves (Figure 150).

3. Ascenders usually start with a serif similar to the beginning of the n (Figure 151). If an inverted arch form is used, it should start at the waistline, which is the upper limit of the x-height (Figure 152). Figure 153 shows wrong forms.

Ascenders with reversed arch forms at the entrance stroke can sometimes be seen. Such forms should be left for more formal modifications of the italic (see Humanist Italic, page 112).

Descenders are considered to start at the middle of the x-height. The letters j and p can be written with an upstroke, even if no connection is possible (Figure 154). Ascenders and descenders can be no longer that those of roman letters, but caution is advised: do not use extreme forms; aim for harmonious ratios.

4. In oval forms consisting of an arch and an inverted arch, such as an o, the upper element should be rounder than the lower one. The lower part can be almost pointed (Figure 156), but it should never look like those in Figure 157.

In the letters a, d, p, g, and q the axis of the oval is more slanted than that of the stems (Figure 158). In a, g, and q, the stem can be drawn slightly higher than the arch (Figure 160).

Up-and downstrokes of v and w show a somewhat modified slant.

All letters are made up of these elements. Practice them often. Start with m, n, h, and u, and add one letter at a time (Figure 163). Connect the letters with each other wherever possible, espe-

Figure 146 *Pen angle*
right   too steep   too flat

Slant
right   too steep   too severe

Figure 147
1. two separate strokes   2. one stroke

Figure 148 *Incorrect forms.*
too pointy   too round   arch broken   too wide   too narrow   loop instead of covering stroke

149 Right   wrong   **150** Right   wrong

Figure 151   **Figure 152**   **153** *Bad beginnings and connections.*

154 *Right.*   **155** *Deformed descenders.*

156 *Right.*   **157** *Wrong.*

Figure 158   **159** *Incorrect forms.*   **Figure 160**

59

abcdefghijklmnop
qrstuvwxyz ßẞ ff

abcdefgkppß adgq

bmondmhpncenam

mozart satin glückl

quousque tandem h

aabbcceeeeeeffggggpphk
kkrrrrsstvvvz

cially at the waistline, to achieve a smooth and flowing appearance. Form only logical connections; do not force them. Figure 166 shows various workable combinations.

Keep comparing your work to a model, and make large versions of particularly troublesome letters at the edge of your sheet as a constant reminder of their correct shape.

161 *Italic.*

162 *Stroke directions.*

Figure 163

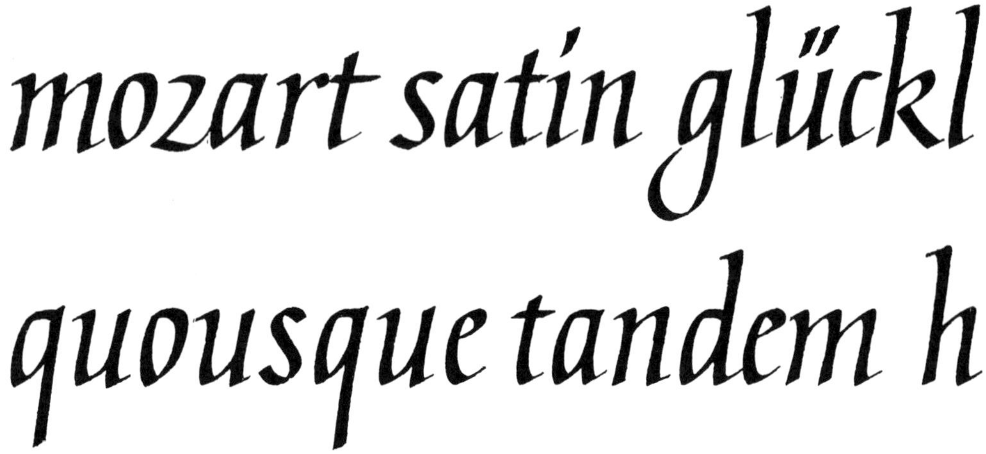

Figure 164

165 *Incorrect forms.*

On page 61:
166 *Some letter sequences and possible connections.*

man ab ag at mbm mbn beben ba

ob obo mcn oco ac ck ck eck odo

oeo ea ed ei es ep ew ex eh eh eb eb ef

fm fe offen affe pfe mgn eg tgo oho

min oio mjn oj el olo als allg mpm

ope ope appe up mqu oque mrn

ors arz irren rg rf msm os ts so tm

es mum ouo ove vn va mxm oxe xt

mym oy ay mzm oze eza tzt zm

Uppercase Letters and Numerals

The proportions of the roman capitals have to be preserved in italic, in spite of the slant. The contrast between the slim lowercase and the comparatively wide uppercase letters adds interest to the lettering. To maintain continuity, do not increase the size of the capitals to more than one and a half times the x-height.

If you have practiced the alphabet shown in Figure 138, a slanted version of it should not be challenging. Remember that the angle of the already slanted downstrokes changes in an overall slanted letter (Figure 169).

Ornamental forms are discussed in chapter 3, but they should not concern the beginner.

Alternate Method

Another road leading to formal italic script begins right with your own handwriting. Start by reducing the frequent connections that result from up and down movements; they tend to obscure the basic forms. Retain only true connections, and gradually change loops into serifs; keep the bottom of the g open.

Using a narrow nib and keeping a small distance between letters will make your work easier (Figure 171). Changes like this will most likely improve every individual letter itself. Pay special attention to the group m, n, h, and u, and ensure a distinct difference between arch and inverted arch forms. Personal handwriting styles often blend both these forms into one. Finally, compare each one of your letters to the alphabet shown in Figure 161 and make changes where they are necessary.

Applications

Variations of italic are plentiful. A close relationship exists between italic and personal handwriting styles, which de-

Figure 167

Figure 168

Figure 169

Figure 170

Figure 171

der Grünspecht kommt zur Nachtigall

der Grünspecht kommt zur Nachtigall

der Grünspecht kommt zur Nachtigall un

der Grünspecht kommt zur Nachtigall u

der Grünspecht kommt zur Nachtigall

172 Development of a formal italic from an individual handwriting style (reconstruction).

der Grünspecht kommt zur Nachtigall und

der Grünspecht kommt zur Nachtigall und

der Grünspecht kommt zur Nachtigall und

der Grünspecht kommt zur Nachtigall und

der Grünspecht kommt zur Nachtigall und

173 Modification of a personal handwriting style through frequent practicing of a formal italic (reconstruction).

veloped from humanist scripts. Information technology and the need for speed endanger good taste and the process of handwriting itself to the point where little is left of the high standard of writing that was achieved during past centuries. What a pity, considering that second only to your language, your handwriting can be the most telling expression of personality! Activities such as speaking and eating are subject to certain rules when pursued in public. Similarly, handwriting should be readable with ease and pleasing from an aesthetic point of view. The advanced state of technology and communication provides us with tapes, data-storage systems, computers, and other machines that free our handwriting from the demands of speed and afford us the leisure of developing a distinct personal script. Every good education should include methods of storing text quickly. Where machines are impractical, stenography or another speed-writing system should be used.

In some parts of the world efforts are underway to halt the deterioration of personal handwriting skills through a reorientation towards the humanist scripts. In Germany, for example, an improved basic alphabet is being taught to schoolchildren to give them cultured writing skills as an integral part of their education.

Good writing tools are essential. Ballpoint pens and felt-tip pens are practical but not suitable. The equal stroke widths they produce lack rhythm and interest. Ballpoint pens make varying pressure impossible; up-and downstrokes look the same. Pressure points appear at illogical spots, and ugly forms result (Figure 174). What should be used instead are wide-nibbed pens. Mature handwriting develops over time; it grows with the personality of the writer. The alphabet in Figure 162 can therefore be no more than a starting point, from which repetition forms a rhythmic sequence that seems to flow from the subconscious mind. Over time individual variations will emerge and should occasionally be compared to basic forms and corrected if necessary.

Figures 175–178 show samples of beautiful handwriting based on humanist scripts.

If you practice italic alphabets with even-stroke pens or a brush, do not use serifs or connect the letters to each other. The resulting even-stroke cursive is well suited for lettering of technical material, plans, maps, signs, or similar uses (Figures 179 to 183).

Uses of the italic for artists are discussed in the section on Humanist Italic, page 112.

174 *Deformed shapes.*

> And there were in the same country shepherds abiding in the field keeping watch over their flock by night. And lo the angel of the Lord came upon them, and the glory of the Lord shone round about them: and they were sore afraid. And the angel said unto them, Fear not: for behold, I bring you good tidings of great joy, which shall be to all people. For unto you is born this day in the city of David a Saviour, which is Christ the Lord.

175 *Italic by Tom Gourdie.*

176 *Jan Tschichold's handwriting. From Renate Tost,* Die Schrift in der Schule: Ein Beitrag zur Schreiberziehung in der allgemeinbildened und polytechnischen Oberschule *(Teaching handwriting in school). Leipzig, 1968.*

177 *Paul Standard's handwriting.*

178 *Example of a good adult German handwriting style. From Renate Tost,* Die Schrift in der Schule.

abcdefghijklm
nopqrstuvwxyz
ABCDEFGHIJK
LMNOPQRSTU
VWXYZ

179 Sans serif even-stroke cursive — German school script. Developed by Renate Tost. From Renate Tost, Die Schrift in der Schule.

180 Example of an application of school script. From Renate Tost, Die Schrift in der Schule.

On page 67:

181 a and b Even-stroke italic. Drawn with a round brush by Harald Brödel.

182 Stroke directions for drawing italic with a round brush. Harald Brödel.

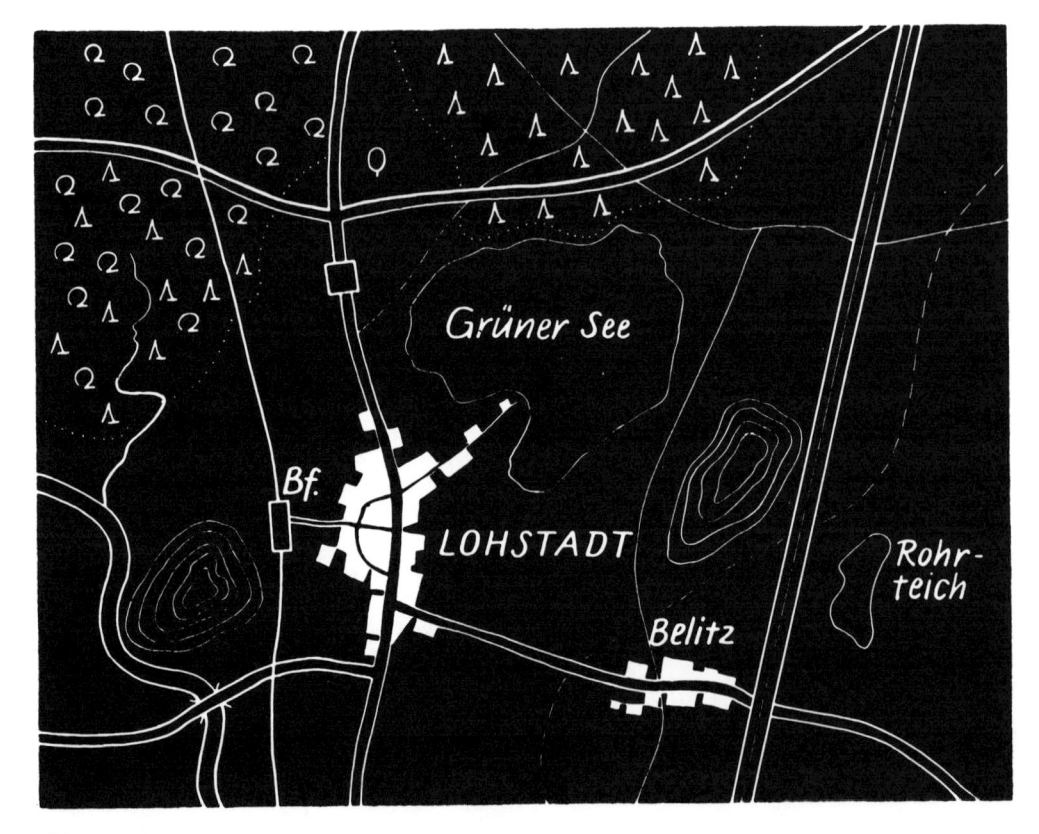

═══ Autobahn
═╡═ Fernverk.straße
═══ Straße 1.Ordnung
─── Straße 2.Ordnung
─── Feld- u.Waldweg
─╼─ Eisenb. mit Bahnhof
- - - Kreisgrenze
〜 Fluß
⌒ Bach
Ω Laubwald
Λ Nadelwald
Ω Λ Mischwald

ABCDEFGHIJKLMNOPQR
STUVWXYZ ÄÖÜ
abcdefghijklmnopqrstuvwxyz
ßäöü (.,,;:!?--/%) 1234567890

Figure 181a

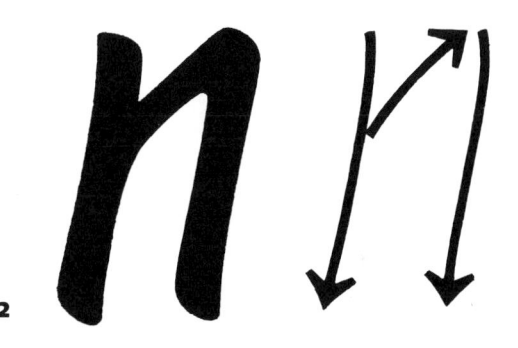

Figure 182

Figure 181b

Die Gleichstrichkursiv ist eine schnell schreibbare, für jedermann leserliche und in ästhetischer Hinsicht den Anforderungen genügende Schrift. Mit Gleichzugfeder oder Rundpinsel geschrieben ist sie für die Beschriftung der Preisschilder geeignet.

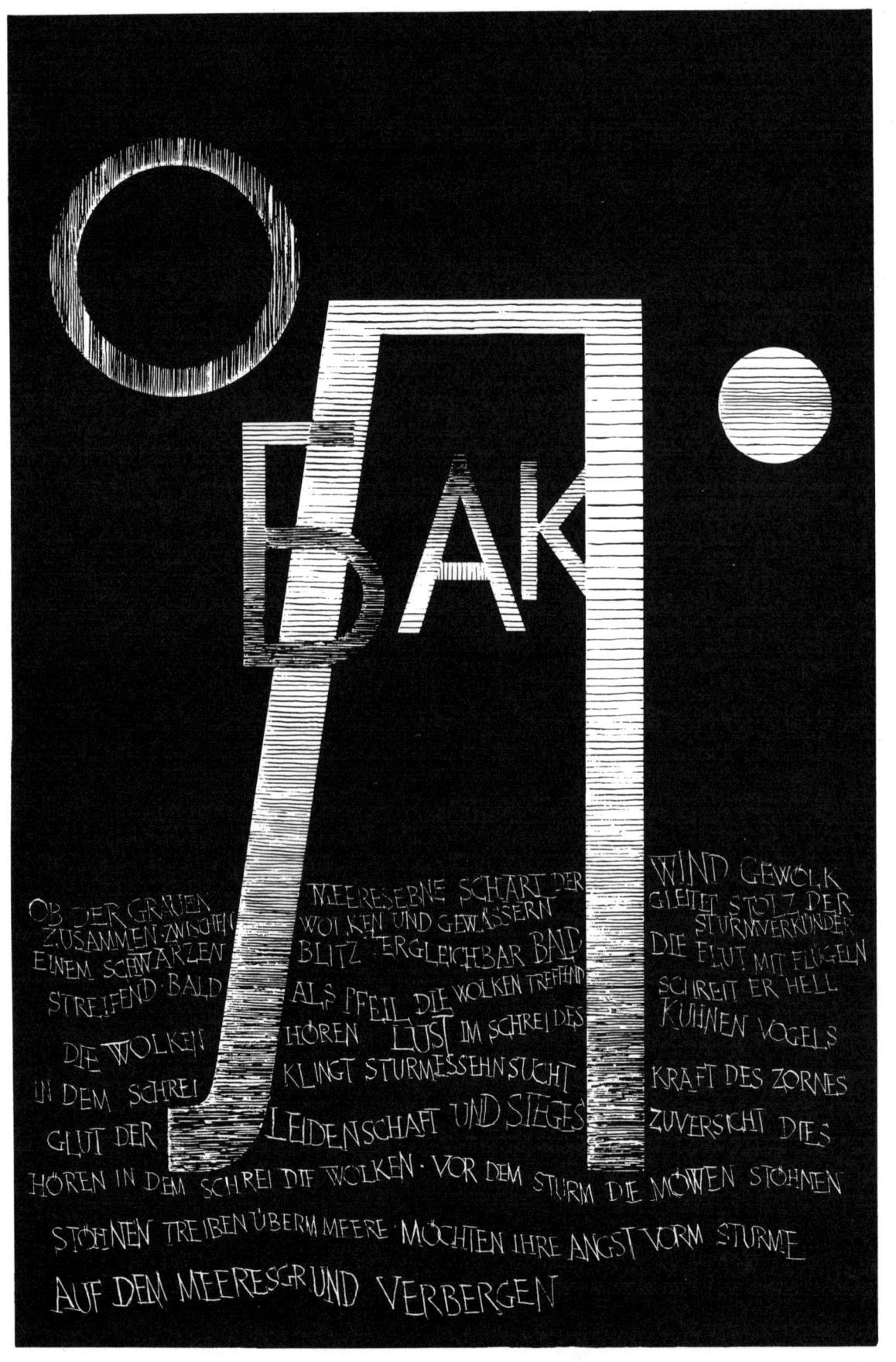

183 *Handlettered page by Irmgard Horlbeck-Kappler from a collection of texts from the rhythmical prose work "Das Lied vom Sturmvogel" (Stormbird's song) by Maxim Gorky.*

CHAPTER 3

Advanced Course

Genealogy of Scripts

ONSVMPTASASOLORESTIT — Roman monumental capitals

- Earlier Roman cursive
- Later Roman cursive
- Uncials
- Half uncials
- West Gothic
- Old Italic
- Merovingian
- Beneventan
- Irish-Anglo-Saxon roundhand
- Irish-Anglo-Saxon
- Carolingian minuscule (ninth century)
- Carolingian minuscule (eleventh century)
- Gothic bookhand
- Renaissance minuscule
- Rotunda
- Textura
- Gothic cursive
- Gothic
- Batarde

The first printing types

- Roman — agricolismū
- Gothic — gaudios
- Rotunda — habensgd
- Textura — pritz luis
- Schwabacher — Rheinlandg
- Fraktur — Reiß yn ortstrich

70

184 *The development of lettering styles from Roman antiquity to the Renaissance.*

INTRODUCTION

"Letters evolved historically. They must be studied to be mastered. No one can invent letters by himself. At best we can modify them."[1]

The full scope of cultural and historical relevance cannot be addressed in the confines of this book. Two works that are intended for the professional present writing as the essence of our culture in its connection to economy and society. They are Albert Kapr's *Schriftkunst* (The Art of Lettering)[2] and Frantisek Muzika's *Die schöne Schrift* (Beautiful Lettering).[3] Only with insight into such connections is a true understanding of changing forms, expressions, and styles possible. In this book only the most basic references are made to the historic relevance of any given style or to materials and tools. The user will find tips on how to avoid the most common mistakes. Illustrations show basic alphabets and possible variations. These are not always intended as models to be copied, but rather as inspiration for the student who wishes to delve deeply into the matter and make historical forms his own, in the sense of the introductory quotation.

Chapter 2 of this book teaches technical drawing skills; this chapter progresses methodically through different writing styles, arranged in historical sequence, because each new development built on the foundations of earlier ones.

1. Jan Tschichold, *Treasury of Alphabets and Lettering*. Reprint. New York: Design Press, 1992. Copyright © 1952, 1965 by Otto Maier Verlag, Ravensburg.
2. Albert Kapr, *Schriftkunst: Geschichte, Anatomie und Schonheit der lateinischen Buchstaben* (The art of lettering: history, anatomy, and aesthetics of roman Letters). Dresden: VEB Verlag der Kunst, 1971.
3. Frantisek Muzika, *Die schöne Schrift* (Beautiful writing). Vols. 1 and 2, Prague: Artia Verlag, 1965.

CLASSICAL ROMAN LETTERING

Roman Monumental Capitals

The geometric forms of the western Greek letters form the basis for Roman capitals: square, rectangle, triangle, and circle. It is a monumental writing style, used for inscriptions and closely tied to architecture in character and proportions. To the end of the first century B.C. Roman monumental writing was a style of even stroke widths, without serifs. These developed with an ever-improving technique of chiseling. In later inscriptions it is frequently evident that serifs are formed in an attempt to duplicate the brush strokes of the underlying design exactly. The contrasting thick and thin lines resulted from the differing angles at which brush or chisel were held. Characteristics caused by materials and tools became design principles. When Roman statesmanship and culture was at its zenith, so was the art of the Roman capital. It was a style in its own right, best exemplified by the inscription at the base of Trajan's Column, which dates from the second century A.D. (Figure 187).

In addition to chiseled inscriptions, lead or bronze letters were riveted to stone. An ornamental variation was achieved here with bifurcated serifs.

Passing through various changes in earlier centuries, the roman capital style saw its renaissance in the fifteenth and sixteenth centuries and is still with us today.

The letters shown in Figures 188 and 189 do not exactly correspond to the letters on Trajan's Column, which does not include the letters H, J, K, U, W, Y, and Z. These have been added to the illustration. There are two variations each of A, M, and N. Many inscriptions show these letters with blunt upper ends

185 *Triumphal arch of Titus, Rome, 81 A.D.*

On page 73:

186 *Fragment of a Roman monumental inscription from the first half of the second century A.D. (From Frantisek Muzika, Die schöne Schrift, vol. 1. Prague: Artia Verlag, 1965.)*

On pages 74 and 75:

187 *Inscription from the base of Trajan's Column, Rome, 113 A.D. (Photo from the Victoria and Albert Museum, London.)*

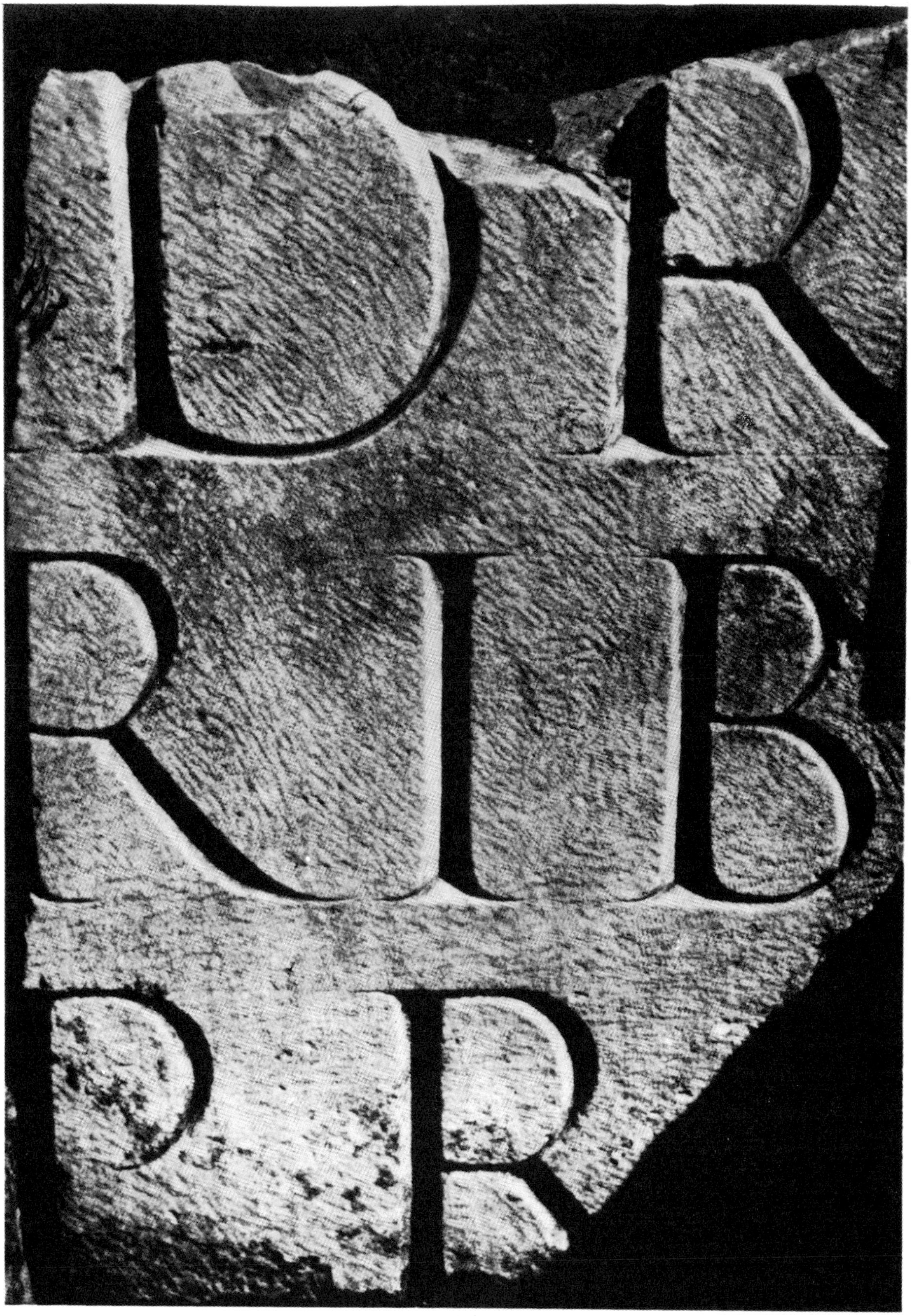

SENATVSPOPVL
IMPCAESARIDIVI
TRAIANOAVGGE
MAXIMOTRIBPOT
ADDECLARANDVMC
MONSETLOCVSTAN

instead of pointed ones (Figure 190).

The following hints will make the learning process easier. Draw each letter on a small card, and arrange them into words. The right stroke widths and proportions can be found more easily this way. Start with H, M, and O to establish size, stroke width and the shape of the serifs. Follow with A, N, U, G, B, R, and S. Only when the appearance of these key figures pleases you should you go on to the rest of the alphabet. Never lose sight of the organic character of your movements. Most of the chiseled inscriptions were originally drawn onto the stone surface with brush and paint.

It is a good idea to practice the letters first with a reed pen in a corresponding

SQVE·ROMANVS
NERVAE·F·NERVAE
RM·DACICO PONTIF
XVII IMP VI COS V PP
VANTAE ALTITVDINIS
IBVS SIT EGESTVS

width. Let the movement come freely from your elbow and shoulder joints. Only your pinky braces your hand against the writing surface. A letter height of 2 inches is a good choice. The ratio of basic stroke width to the letter height is 1:10 in the middle and 1:9 in the upper and the lower part of the shaft (Figure 191). Small adjustments of these measurements can be made according to personal preference or to accommodate special conditions of the work. Make the horizontal strokes half the width of the basic stroke, the thin diagonal strokes about two-thirds of it. Figure 188 offers help with the construction of the letters. One unit corresponds to one-tenth of the letter height.

188

(H) Use a width of about 8 units. Draw the crossbar on the middle line.

(T) The crossbar is 8 units wide. Its thickness is two-thirds that of the vertical stem. The angles of the serifs are slightly different.

(E) The width of the upper crossbar is 4¾ units; the middle bar is somewhat shorter, the bottom bar somewhat longer and wider than the upper bar. The bottom bar ends in a slanted serif.

(F) Draw the F somewhat wider than the E. Place the second crossbar lower than that of the E.

(L) Make the crossbar of the L 5⅓ units wide. The letter looks more beautiful if it is wider than the E.

(A) The base is 8 units wide. The crossbar is half the width of a the basic stroke.

(N) The N is 8 units wide. The width of the vertical lines is three-quarters that of the basic stroke.

(W) The W consists of two connected V-shapes, but the interior spaces are somewhat narrower. It is also possible to create the W from interlocking Vs. In this case the interior space of the Vs is not diminished, but the slant of the inner diagonals has to be flatter than that of the outer ones to ensure that the middle triangle is not too small.

(M) The width of the M at its base is 11 units. The first and last diagonals are less slanted than the middle ones. The slant of the angle amounts to approximately one basic stroke width. The first and last diagonal have two-thirds the width of a basic stroke.

(K) The K may be drawn with a straight downstroke or with one that is slightly curved, like that of an R. Use two-thirds of the basic stroke width for the upper diagonal. Both diagonals meet in the optical center. Draw the lower part of the letter one unit wider than the upper part.

(X) The upper width of the X is about 7 units. The thinner diagonal is half as wide as the basic stroke.

(Y) The width of the Y is 7½ to 8 units. The diagonals meet at the geometric middle, so the interior space is larger than the space inside the upper part of the X.

(Z) A width of 8 units is appropriate. Draw the horizontal lines half as thick as the diagonal, and the lower serif more slanted than the upper.

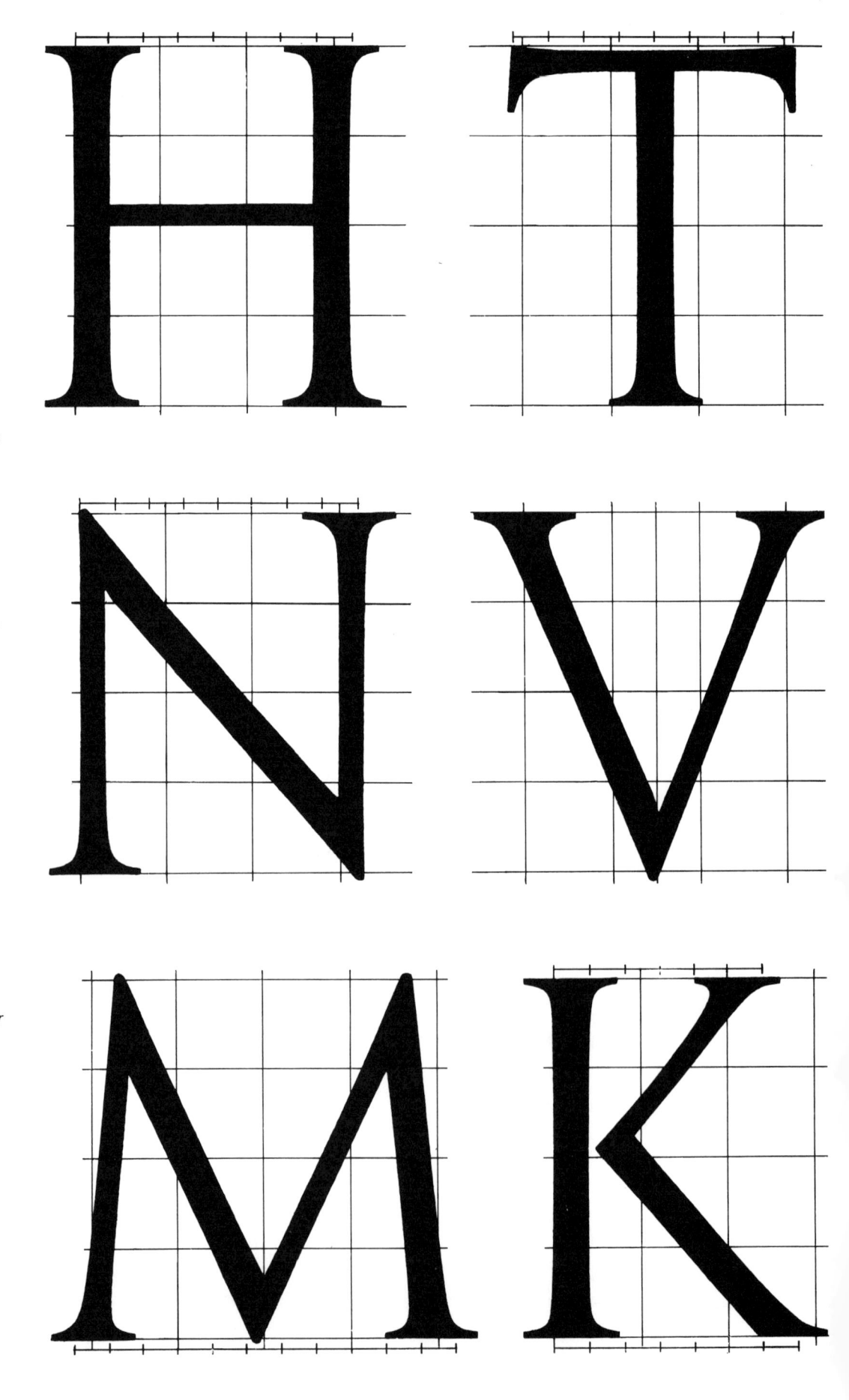

77

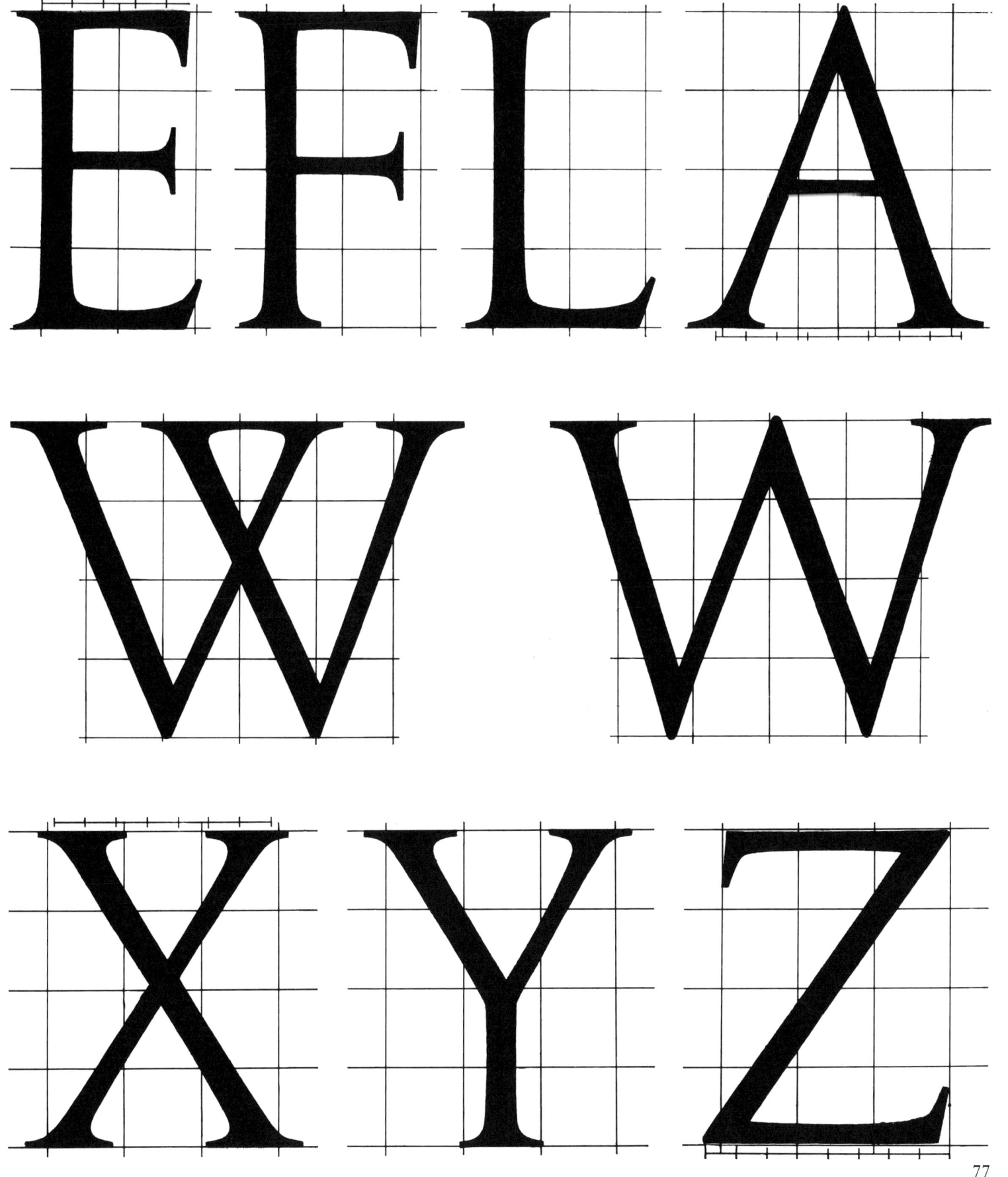

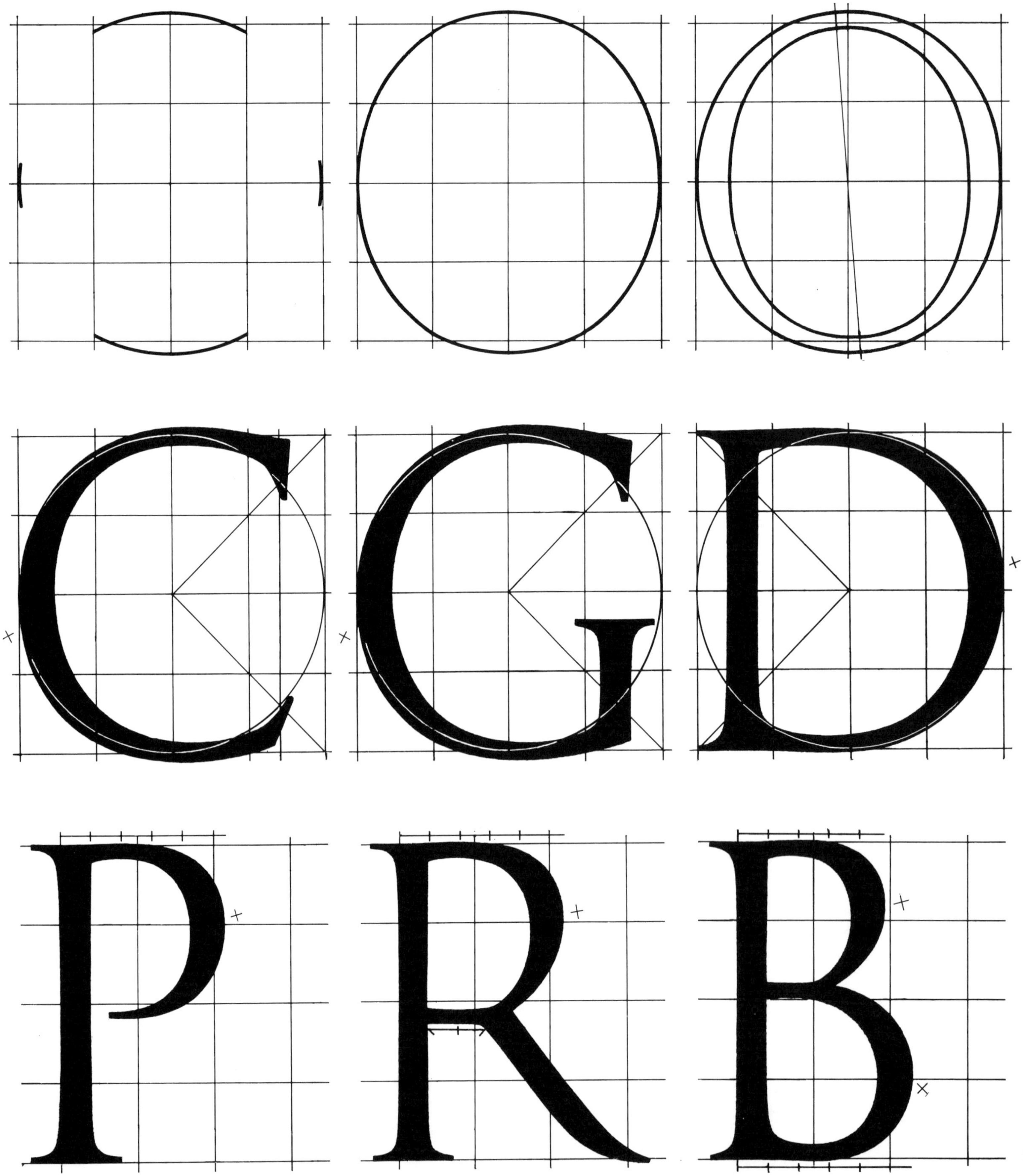

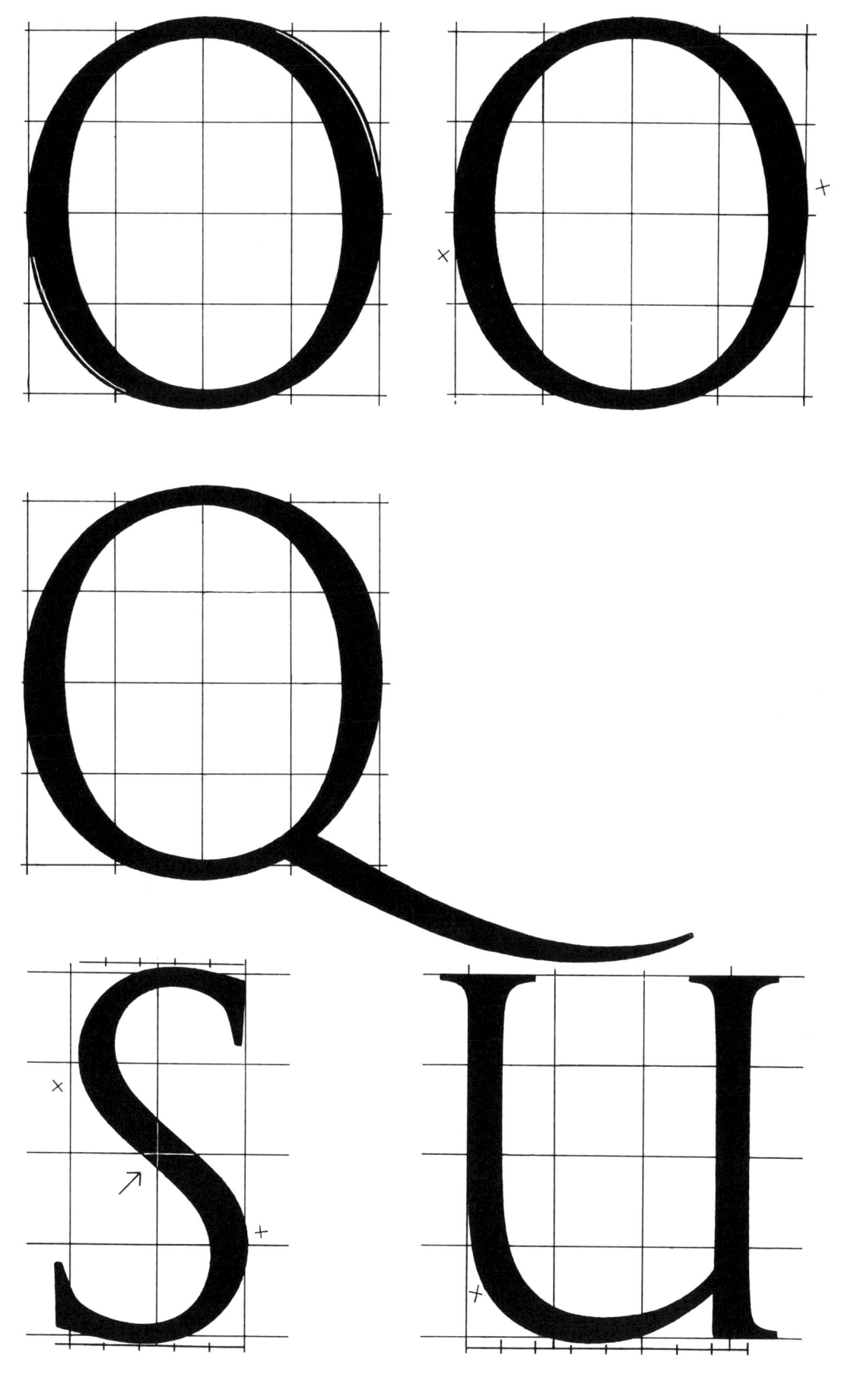

(O) Let the upper and lower curves of the O extend slightly beyond the guidelines to make this letter appear the same size as the others. The contour looks almost like a circle, but the oval that defines the inner space has a slanted axis. The arch is marked with an x on the parts where the stroke width is even thicker than that of the basic stroke. The thinnest parts correspond to half the basic stroke width.

(C, G, D) These letters are about 9 units wide. The width can also be constructed using the inscribed circle and the diagonals. All forms are derived from the letter O, but the curves on top and bottom are slightly flatter. The G on Trajan's Column has a vertical bar that is 4½ units long, but it is acceptable to make it shorter. The serifs on the C are smaller and more slanted.

(P) The P measures 5½ units on its widest part. The arch reaches below the geometric middle and remains open at the bottom. To give the arch the same weight as the bar, draw it thicker at the spot marked with an x. The same is true for the arches of R and B.

(R) The upper part is 5½ units wide. The middle horizontal lies below the geometric middle and is thinner than the upper horizontal. Start the downstroke at a distance of 1¾ units from the vertical bar and carry it below the baseline to ensure that it will not look shorter than the bar.

(B) The greatest width in the upper part of the B is 4¾ units; in the lower part, somewhat less than 6 units. Place the separation in the optical middle. Draw the horizontals less than half the width of the vertical bar. Keep the middle horizontal thinner than the upper one, but the bottom horizontal line should be slightly thicker.

(S) The upper part of the S is 4¾ units wide, the lower part 5½. It is constructed from two semicircles that are flattened on top and bottom. The best inscriptions contain S shapes that seem as if they were tipping forward slightly.

(U) The U is 8 units wide. For optical reasons the arch reaches below the baseline. The widest part is marked with an x.

ABCDE FGHJ KLMN

189 *Drawn roman capitals. Study by the author.*

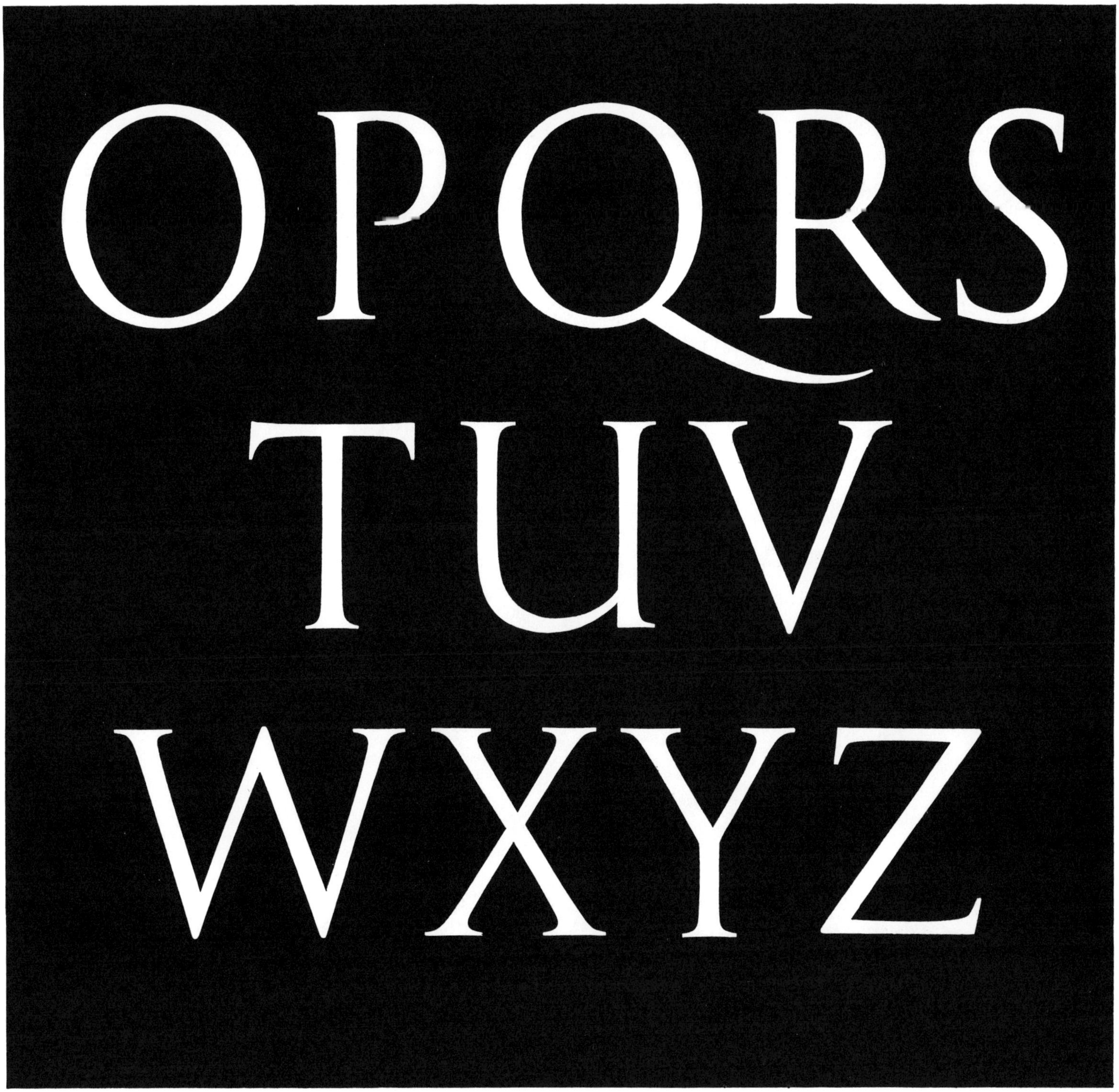

OPQRS
TUV
WXYZ

Figure 190
Figure 191 AMN I →← 1:9 
　　　　　　　 →I← 1:10
　　　　　　　 →I← 1:9

Rustica

We use the term *rustica*, even though it is controversial in literature. Rustic capitals resulted from a calligraphic modification of Roman monumental capitals. These were used to call the Romans to the polls, for official announcements and news items, often written on walls in red and black ink. Some examples are preserved in stone or bronze. Monumental and rustic capitals were often mixed in a single inscription. Rustic capitals were used extensively in manuscripts up to the sixth century, and their form remained largely unchanged. During the eleventh century the use of rustic capitals was limited to headings.

The basic forms of the rustic capitals are not the square and the circle, but the rectangle and the oval. Narrower variations followed wider ones. Stroke variations can be achieved with a wide brush: verticals appear rather thin, horizontals including serifs are wide. The pen angle is usually slanted. The nib is almost held at a right angle to the baseline, which creates the thin vertical and wide horizontal strokes. For practice use reed pens, steel nibs, or flat brushes.

192 *Detail of an inscription on a Pompeiian wall. (Photo by Albert Kapr.)*

193 *Rustic capitals (detail). Handlettered inscription from the fifteenth century. From Chatelain,* Paléographie des classiques latins.

JONAS KAPITOL HEBRON

MIDIAS BRASIL CAESAR

DAVID RUSTIKA URBAN

NASE ORNAT QUINTUS

LOTOS WACHTEL XAVER

YSOP ZASTER PHÖNIX

ΓΓΛ
ΙΖΟ

ICH HALTE DIEJENIGEN FÜR GLÜCK-
LICH, DENEN ES VON DEN GÖTTERN
VERGÖNNT WORDEN IST, ETWAS ZU
LEISTEN, WAS AUFGESCHRIEBEN ZU
WERDEN, ODER ETWAS ZU SCHREIBEN,
WAS GELESEN ZU WERDEN VERDIENT.

194 *Pen angle.*
195 *Rustic capitals (detail). Study by the author.*
196 *Rustic capitals (detail). Study by Harald Brödel.*

Quadrata

Quadrata, or square capitals, were the other form of lettering used for manuscripts in the ancient world. Muzika refers to new research when he call them "a paraphrase of Roman monumental capitals." Parchment fragments can be dated back to the fourth and fifth centuries, and the style was used for headings up to the eleventh century.

Square capitals are not constructed and should present no special difficulties, since the basic forms are practically identical to those of the sans serif roman of different stroke widths discussed in Chapter 2. The only difference is the angle at which the pen is held, which is flatter.

Earlier and Later Roman Cursives

Epigraphs and books were lettered in formal scripts. Almost all of them show variations in handwriting style which are of great importance to the development of lettering as a whole. All cursive styles belong in this category and were developed in response to the need for speed in writing. The resulting changes of basic forms include: simplification of basic shapes, connection of letters with each other, ligatures, and, on occasion, slanted letters. To maintain legibility as letters became more uniform, ascenders and descenders were developed.

The earlier Roman script was in daily use for manuscripts throughout antiquity. The simplified letters of the monumental Roman capitals were scratched into wax tablets with a stylus or onto papyrus with a calamus. By the fourth century A.D. the Roman cursive style had developed into a full alphabet of minuscules with ascenders and descenders. Different national variations of it existed until the time of the late Carolingians.

197 Quadrata (historic form). From Arndt/Tangl, Schrifttafeln zur erlernung der lateinischen Paläographie *(Instructional tables of latin paleography)*. Berlin, 1904.

198 Earlier roman cursive (historic form). From Brückner/Marichal, Chartae latinae Antiquiores. Olten and Lausanne, 1963.

199 Later roman cursive (historic form). From Brückner/Marichal, Chartae latinae Antiquiores. Olten and Lausanne, 1963.

200 Pen angle and biassed stress.

201 Quadrata. Study by the author.

Figure 197

Figure 198

Figure 199

Figure 200

Figure 201

202 *Uncials. From Erik Lindegran, Vara bokstaver. Goteborg, 1959/60.*

Uncials and Half Uncials

Uncials resulted from the development of the early Roman cursive into a bookhand. It contains elements of both the cursive and the classical capitals.

Some examples suggest Greek uncials as models; the A and the rounded M especially seem to come directly from Greek sources. Shafts of the D, F, H, L, P, and R developed into ascenders or descenders. The substitution of curves for straight lines can be attributed to Byzantine influences in culture and architecture. The final form of the letters was made possible by the fact that parchment replaced papyrus and quill, which in turn had replaced the stylus. All uncial styles show a strong thick-thin contrast. Up to the sixth century a flat angle of the pen or a differently cut nib produced an almost vertical stress. The shafts had a blunt beginning and acquired slanted and later horizontal entrance and exit strokes over time. During the early Middle Ages headings and first letters of lines or paragraphs were created from composite forms: outlines were drawn first and filled in later.

Up to the ninth century half uncials were in use together with uncials. The half uncials were a bookhand developed from elements of the Roman capitals, and of the earlier and later Roman scripts, the latter of which contributed a large number of shapes. The later Roman script also influenced the development of descenders and ascenders, which typically are shaped like a club. Strong differentiation between thick and thin strokes is a similarity to uncials, and so is the emphasis on rounded forms, but the axis is slanted towards the diagonal, giving a biassed stress.

Early medieval book and papal hands are based on form elements of the earlier Roman script. European national styles like the West Gothic and the Merovingian show influences of half uncials, whereas the Irish and Anglo-Saxon styles seem to be variations of the half uncials. All of these forms developed characteristic eccentricities or decorative elements to a point where further development was precluded.

203 *Uncials (historic form).*

204 *Half uncials (historic form).*

On page 87:

205 *Uncials. Detail of a study by Harro Schneider.*

206 *Uncials. Study by Albert Kapr.*

207 *Half uncials. Detail of a study by the author.*

208 *Half uncials. Detail of a study by Harro Schneider.*

209 *Ductus.*

210 *Variation of uncials. Study by the author.*

211 *Variation of uncials. Study by the author.*

212 *Half uncials. Study by the author.*

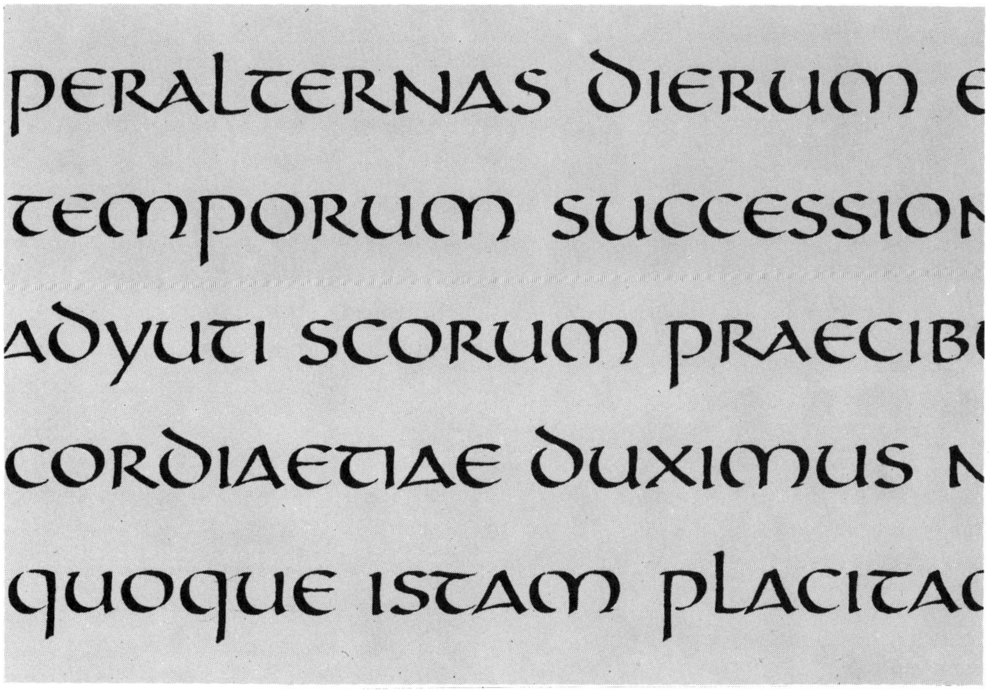

peralternas dierum e
temporum succession
adyuti scorum praecib
cordiaetiae duximus
quoque istam placitac

abcdefGhiklmnopQrts

einstige stürme verstreut
lasst uns alles lebendige
korn zusammentragen. ich

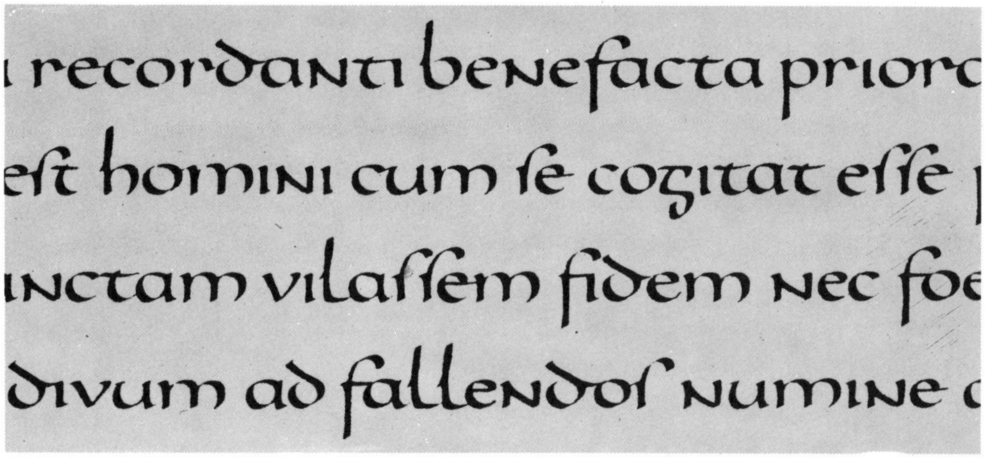

recordanti benefacta prior
eft homini cum se cogitat esse
nctam vilassem fidem nec foe
divum ad fallendos numine

OMEB

Figure 209

ABC
DEFG
HIJK
LMNO
PQS

Figure 210

ABCDEF
GHKL
MNOPTQ
RSU

Figure 211

Figure 212

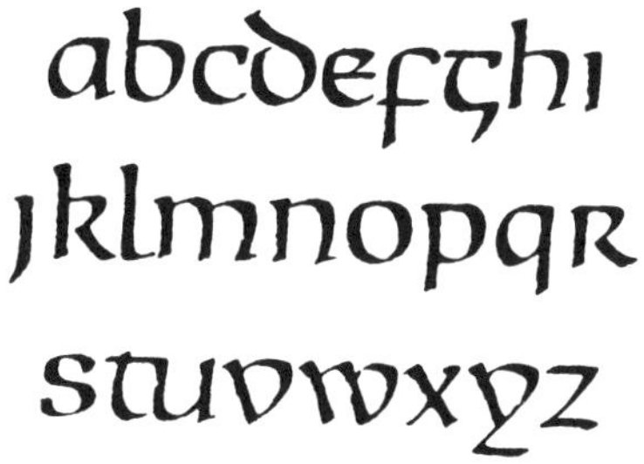

abcdefghi
jklmnopqr
stuvwxyz

> indulcedine tua.
> pauperi ds̄.
> Dn̄s dabit uerbum
> euangelizantibus:
> uirtutem multam.
> Rex uirtutum. di
> lecti dilecti. et
> speciei domus. diui
> dere spolia
> Si dormiatis inter

213 *Carolingian minuscules (historic form). From Anton Chroust,* Monumenta Palaeographica. *Leipzig, 1915.*

214 *Carolingian minuscules of the twelfth century (historic form). From Anton Chroust,* Monumenta Palaeographica.

> tione susceptum. intellectus
> primo diuine pagine leuiori
> dam estimaui. Mox itaq; ut
> legitur: inueni in illo decuplum
> ligentia sup coeuos eius. Mirum
> dum tenera etas celesti irradia

Carolingian Minuscule

The Carolingian minuscule almost concludes the development from majuscule (capital) to minuscule (small letter, or lowercase). It was developed from half uncials and still-dynamic elements of contemporary scripts and bookhands of Frankish chanceries, and instituted as the official script by Charlemagne around the year 800. Similar tendencies could be observed outside his empire. The ultimate goal of a reform in handwriting was the institution of Christianity in Europe and the raising of the general educational level, which was unthinkable without studying the cultures of antiquity, according to Muzika. In spite of all its advantages, the new alphabet was slow in replacing the many regional west European scripts.

Carolingian minuscules are known for their clear forms. They are easy to read because of their well-developed ascenders and descenders, clear entrance and exit, and the contrast between thick and thin lines. They remained in constant use for five centuries. The remaining examples are characterized by wide shapes within the x-height, flat arches, round forms that are almost circles, a biassed stress, and long, emphasized ascenders with a slight slant. The ninth and tenth centuries brought a refinement of graphic details such as entrance strokes for ascenders that resemble serifs and elongated stems of the letter a. Further minor variations appeared at the end of the tenth century and especially during the eleventh century. Uncials, rustics, and roman capitals were drawn with a quill, and filled in with color.

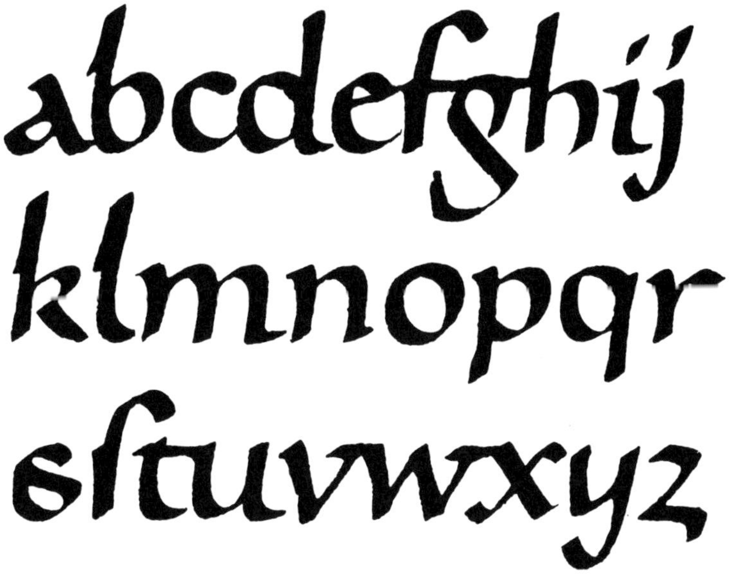

abcdefghij
klmnopqr
sſtuvwxyz

Figure 215

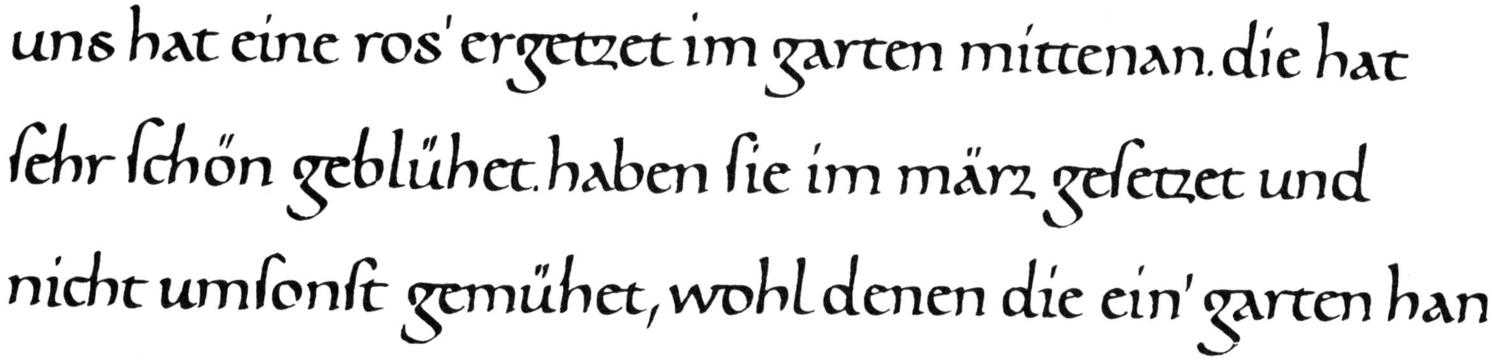

uns hat eine ros' ergetzet im garten mittenan. die hat

ſehr ſchön geblühet. haben ſie im märz geſetzet und

nicht umſonſt gemühet, wohl denen die ein' garten han

Figure 216

Figure 217

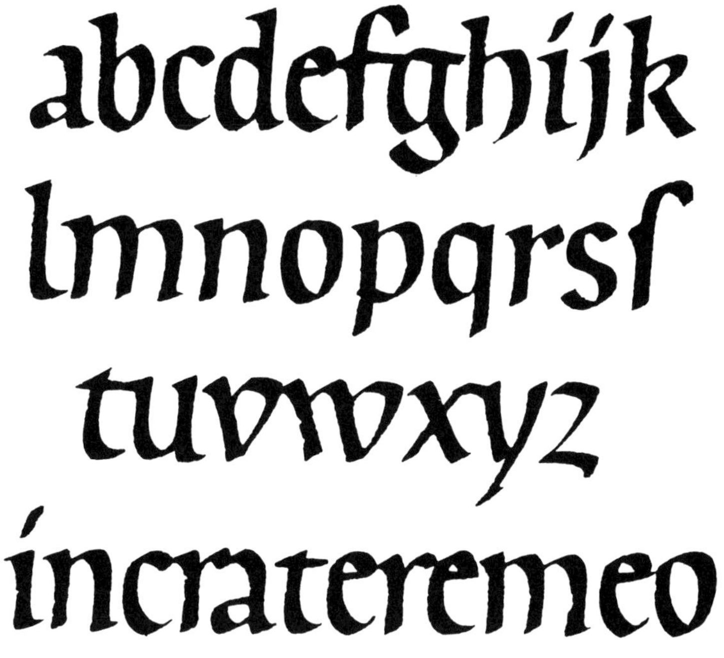

abcdefghijk
lmnopqrſ
tuvwxyz
incrateremeo

215 Carolingian minuscules, lettered by the author.

216 Carolingian minuscules, lettered by the author.

217 Carolingian minuscules, lettered by the author.

GOTHIC LETTERS AND THE RENAISSANCE

Textura

Under the influence of the Gothic style in the twelfth century the x-height forms of the Carolingian minuscule grew narrower and taller. The arches, entries, and exits of letters started showing a tendency to "fracture." From these early Gothic bookhands all later textura, or blackletter, styles were generated, by replacing all rounded forms with broken lines. The form canon was established in northern French monasteries at the end of the twelfth century and was gradually accepted by all other western European monasteries. In the German realm textura found an important expression in missals.

Textura letterforms are narrow and written with a broad flat-edged nib for a strong contrast between thick and thin. Verticals are connected to each other with rhomboid forms, and these interior spaces are frequently narrower than the stroke width. Distances between lines and words are very small. The letter a consists of two parts, and a broken uncial form serves as d. The f and the long version of s reach only to the baseline. Legibility is hindered by the dense lettering, by relatively few ascenders and descenders, and by the multitude of ligatures.

Referring to different stem shapes, the following groups can be distinguished:

1. Textura with rounded ends (fourteenth century)
2. Textura with broken ends (fourteenth century)
3. Textura with straight ends (fourteenth to fifteenth centuries)
4. Textura with rhomboid-shaped ends. This is a late development, which served as the model for the type used in Gutenberg's Bible.

The third form of blackletter appears in French manuscripts. A skillful twist of the pen in the lower part of the stem creates horizontal ends. Since fractured forms appear only at the upper but not at the lower ends of letters, a very attractive contrast results.

A fourth variation is the execution of textura in stone, bronze, or wood. Here the letters are not engraved but raised. Some inscriptions were painted or cut into wood for printing purposes. The latter technique furthered ever pointier entrance and exit strokes, an effect that was later mimicked in some typefaces.

For special occasions, forms of roman capitals and uncials as well as rustic capitals with elements of Gothic script were used. An original and very ornamental set of capital letters was developed from Gothic elements and the addition of minuscule forms. These so called Lombard versals were drawn with a brush in attractive contrast to the common style.

Rotunda

Textura could not hold its own for long in Italy and Spain. At the end of the fourteenth century it was fused with Latin traditions into rotunda. Like textura, rotunda is characterized by thick vertical strokes and short ascenders and descenders, but it appears wider, rounder, and lighter in spite of many ligatures and the frequent contraction of letters. Intermediate forms came from encounters with humanist minuscules.

Many typefaces were created in the image of rotunda and a later form of it, the Gothic. They were very legible and popular even outside of Italy. Only one form of it is still in use, in German-speaking countries — Koch's Wallau.

On page 91:

219 Early Gothic bookhand (historic form). From Hermann Degering, Die Schrift (Lettering). Berlin, 1929.

220 Textura (historic form). From Hermann Degering, Die Schrift.

221 Textura (historic form).

222 Textura. The bars have horizontal ends at the baseline (historic form). From Albert Kapr, Deutsche Schriftkunst (The art of German lettering). Dresden, 1955.

223 Textura (historic form).

224 Type from the forty-two line Gutenberg Bible. Mainz, 1455.

225 Stroke endings.

226 Ductus.

227 Woodcut for the title page of Hartmann Schedel's Chronik. Nuremberg: A. Koberger, 1493. From Hermann Degering, Die Schrift.

228 Italian rotunda, fourteenth century. From Ernst Crous/Joachim Kirchner, Die gotischen Schriftarten (The art of Gothic lettering). Leipzig, 1928.

218 Medieval initial. From Graphik, reproduced by kind permission of Karl Thiemig KG, Munich.

uis extiterint aliqui qui ñ
dixerint fuisse cõpositu. se

Figure 219

agmũbz exprobrantẽ. Sper
mus precedentẽ adolescentuli

Figure 220

rum prodigijs exaltauit
pro ut infra plenius inno
tescet · Nam ad dei gloria

Figure 221

quoniam in trauem
que; usq2 ad anima

Figure 222

Figure 225

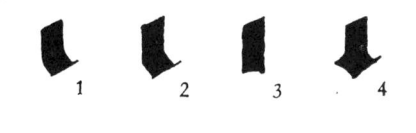

Figure 226

sitatib: sup quibus
antiq; iuris: gratiis
cõmunitatib: esuetu
dinib: seu reb: aliis
etiam proprio motu se
u alias an nobis vel
predecessoribz memorie
diuis romanor im
peratorib: predecessor
ib: nris sub quibz eti
q; vl oriun tenorib:
ecessa t eccesse seu a
nobis vel successor
ib: nris romani imp

Figure 223

Figure 227

udisset dauid: descendit in
philistijm autem venientes
t in valle raphaim. Et cõ
id dũm dicens. Si ascendã
im. et si dabis eos ĩ manu
xit dũs ad dauid. Ascende:
dabo philistijm in manu
ergo dauid ad baalphara
sit eos ibi et dixit. Diuisit
eos meos corã me: sicut di
ue. Propterea vocatũ ē no
i9 baalpharasim. Et reliq
lptilia sua: q̃ tulit dauid et
addiderunt adhuc philisti
dcent: et diffussi sũt ĩ valle
Cõsuluit autẽ dauid dũm.
i cõtra philisteos: t tradas

Figure 224

Figure 228

strauit se in tram p̃tinus
p̃tum potat clamans. Do
im ut sim dignus, q̃ sub t
untres. Meruit hoc peccõr
nõ sum dignus. Numqd
n. q̃ ois pares mei: Tu
i uno ictu oculi te mõstrac.
huihas ut patians ad hoi
publicanũ t peccõre. non
nanducare uis. si te ipm
llo uibes. Cumq3 ipe ul
os engens se uir glosus t
um cunctis tenentibus. ma
t suspirijs. et q̃ plices pau
uum dicit. Tu es deus

abcdefghijklmnopqrsstuvwxyz

229 *Textura. Study by the author.*

Adam bertil cæsar david elbe faust medok niklas ornat jonas kagor pech quintus sebald thüringen vogesen urban wodka xerxes ysop za Hamburg

230 *Textura. Study by the author.*

A·B·K·D·E·S·L·T·N·O·X·E

231 *Lombard versals. Study by the author.*

Das Schwere ist der Kampf, aber der Kampf ist das Vergnügen. Romain Rolland

232 *Textura. Study by the author.*

Rose, o reiner Widerspruch, Lust, niemandes Schlaf zu sein unter soviel Lidern.

233 *Textura. Study by the author.*

234 *Textura. Study by the author.*

abcdefghijklmnopqrsstußvwxyz

235 *Rotunda capitals. Study by the author.*
236 *Textura capitals. Study by the author.*

Figure 237

Figure 238

Zum Besten der gesamten Menschheit kann niemand beitragen der nicht aus sich selbst macht, was aus ihm werden kann und soll.

Figure 239

ABCDEFGHIJKLMNOPQR
STUVWXYZ   1234567890
abcdefghijklmnopqrstuvwxyz

On page 94:

237 *Textura. Study by Albert Kapr.*

238 *Rotunda. Study by the author.*

239 *Caslon Gothic (36-point). D. Stempel AG, Frankfurt/Main. Typesetting by Andersen Nexo, Leipzig.*

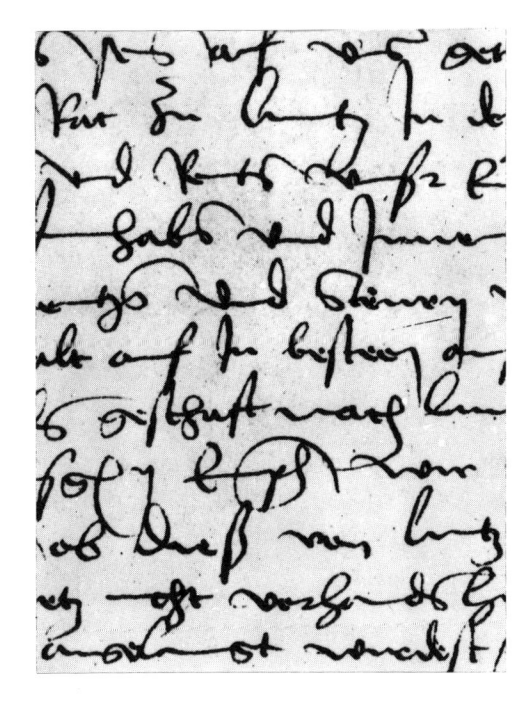

240 *Late medieval cursive (1469). From Chroust,* Monumenta Palaeographica.

241 *French batarde, fifteenth century. From Crous/Kirchner,* Die gotischen Schriftarten.

242 *The "elephant trunk" style. Study by the author.*

243 *Italian batarde (historic form). From Crous/Kirchner,* Die gotischen Schriftarten.

244 *Frankish batarde (historic form). From Crous/Kirchner,* Die gotischen Schriftarten.

245 *Exercise.*

Gothic Scripts and Variations

Gothic Cursive

Textura remained reserved for liturgical use because it was hard to read and slow to write. From the Carolingian minuscule a cursive hand for a wider use was developed. Its characteristic is a high degree of connection between strokes. A fluid writing sequence softened the fractured forms, x-heights were low, ascenders and descenders long. Arches and loops were attached to ascenders, the letters were written with a wide quill, and there was much room for individual variations.

Batarde

Batarde is a group of fifteenth-century styles that are neither true cursives nor bookhands, according to Muzika. From its origins in France and the royal Bohemian chanceries it spread across the German states as far as Poland, the Netherlands, and England. Batarde scripts exist in many variations and served as vehicle for uncounted examples of works of national literature and languages. Typefaces were created in its image and combined with textura capitals. Separate capital alphabets were also in existence. Only the most important variations will be discussed here.

The French batarde is very decorative and elegant. The hour books in the Flemish tradition of illuminated manuscripts are of special beauty. The well-known Civilité, a typeface of the sixteenth century, also has its origin in the French batarde (Figure 241).

A peculiarity of the Bohemian batarde is the so-called elephant's trunks (Figure 242) and the fact that letters are written separate from each other in books. Muzika makes the following distinctions among the Bohemian batardes: round batarde (fourteenth to fifteenth centuries), fractured batarde (fifteenth to sixteenth centuries), and spiked batarde (fifteenth to sixteenth centuries). The Bohemian batarde was of great importance for the development of fraktur.

Wonderful examples of Italian batarde are to be found in the books of Tagliente and Palatino. Italian modifications influenced writers from Spain to the Netherlands. Among the German batardes those from the Upper Rhine and Franconia deserve special mention. The earlier shows a relationship to the French batarde; the latter presents a particularly mature type, which is the basis for other German scripts (Figure 244).

To achieve the flowing rhythm so typical of the batarde scripts, practice the group m-n-u and add one letter after the other (Figure 245). Study historical models carefully before you attempt any variations of your own. Copying is a useful activity, and it is important to assimilate a rich supply of historical information into one's own repertoire.

246 *Batarde. Study by the author.*
247 *Batarde capitals. Study by the author.*
248 *French batarde. Study by the author.*
249 *Batarde. Study by the author.*

On page 97:
250 *Batarde. Study by Alfred Kapr.*
251 *Batarde. Study by Alfred Kapr.*
252 *Batarde. Study by Alfred Kapr.*
253 *Batarde. Study by the author.*

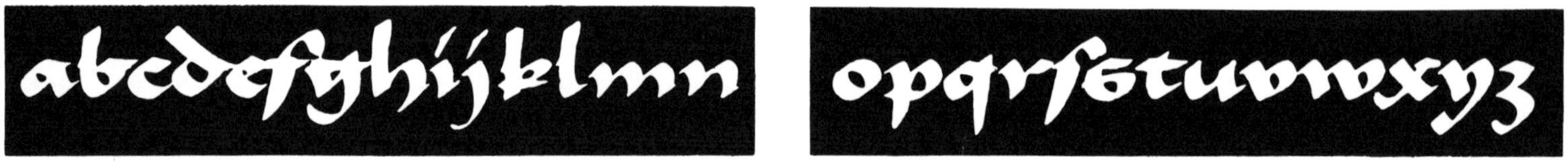

Figure 246

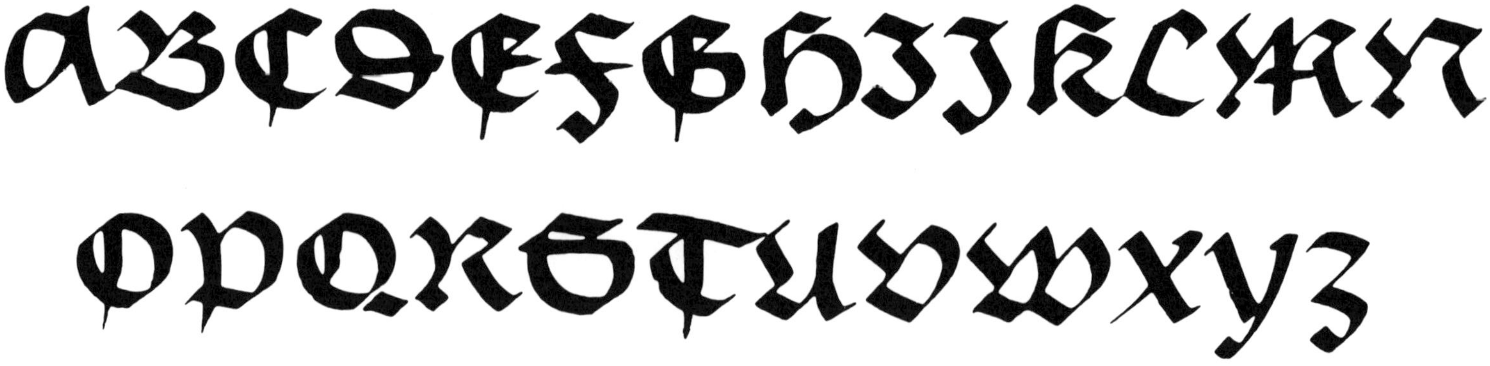

Figure 247

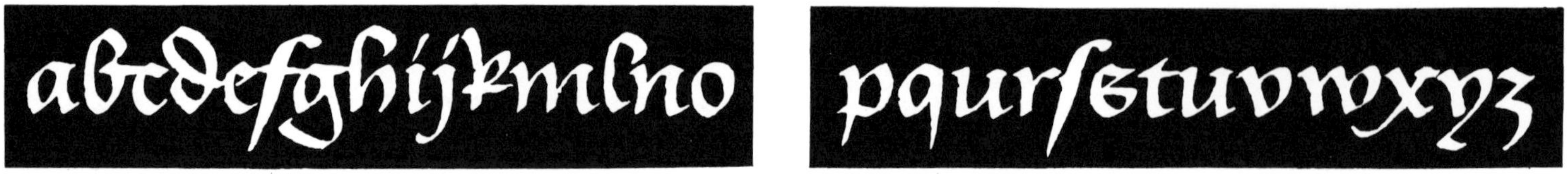

Figure 248

Figure 249

körper und stimme verleiht die Schrift
dem stummen Gedanken
durch der Jahrhunderte Strom trägt ihn
das redende Blatt

Figure 250

Figure 251

abcdefg

und die echte Sehnsucht muß stets produktiv sein, ein neues besseres erschaffen.

Figure 252

Figure 253

meine tinte ist gefroren und ausgeloschen der kamin

Schwabacher

These are the letters of the Reformation movement of the sixteenth century. They are early German batarde forms cut in wood for block prints, which appeared as typefaces from the middle of the fifteenth century.

The origin of Schwabacher is the Frankish batarde, and it shows clear evidence of its beginning in woodcuts. Strong lines and changes between round and straight forms make it interesting and legible. There are clear-cut ascenders and descenders, few ligatures, and a highly original assembly of capitals.

Use a wide reed or quill, and take a typeface specimen as a model, but leave room for changes necessitated by your pen.

Related to the Schwabacher but less expressive are the Upper Rhine and Wittenberg typefaces, both evolved from the Upper Rhine batarde.

Fraktur

A mixture of book textura, batarde, and Schwabacher, fraktur is very decorative and heavily influenced by Bohemian chanceries. The minuscules are dense and appear more precise and differentiated than those of the Schwabacher. Forked ascenders are typical. The capital alphabet is modeled after the form of the Schwabacher capitals, but it is augmented with decorative extension strokes that resemble elephants' trunks, and refined.

Great masters of fraktur were Leonhard Wagner (1457–1522) and Johann Neudörffer the Elder (1497–1563). Fraktur was often cut in wood for use in book titling.

Some of the fraktur types that were in use during the Renaissance served for special documents, and later for increasingly widespread uses. Only some of them can be mentioned here.

254 *Old Schwabacher (36-point). Genzsch & Heyse, Hamburg. Typesetting by Andersen Nexö, Leipzig.*

255 *Variation of Civilité (reduced) by Hermann Zapf. From* Pen and Graver. Alphabets & Pages of Calligraphy. *New York, 1952.*

256 *Ductus for Schwabacher.*

257 *Schwabacher. Study by the author.*

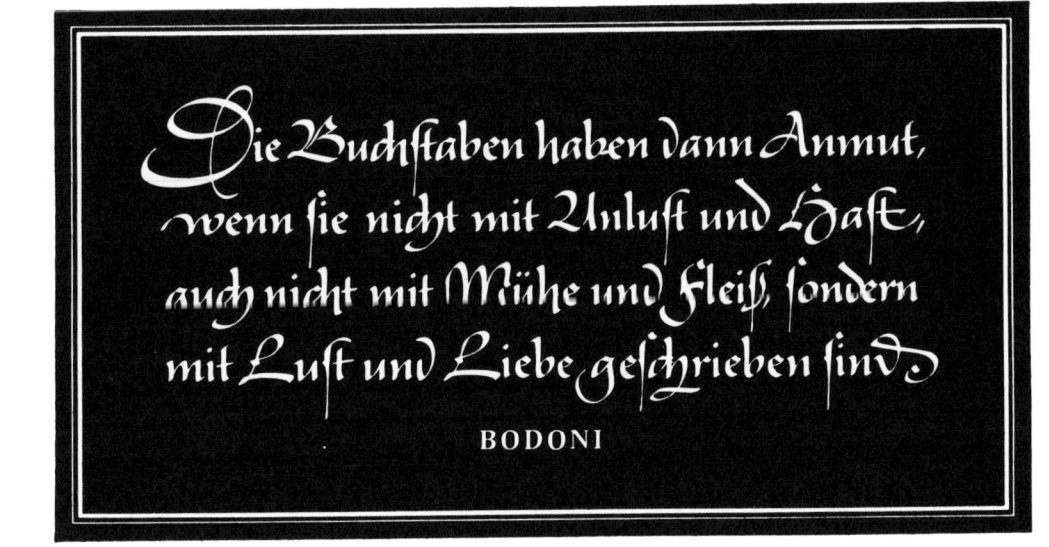

Figure 255

Figure 256

Figure 257

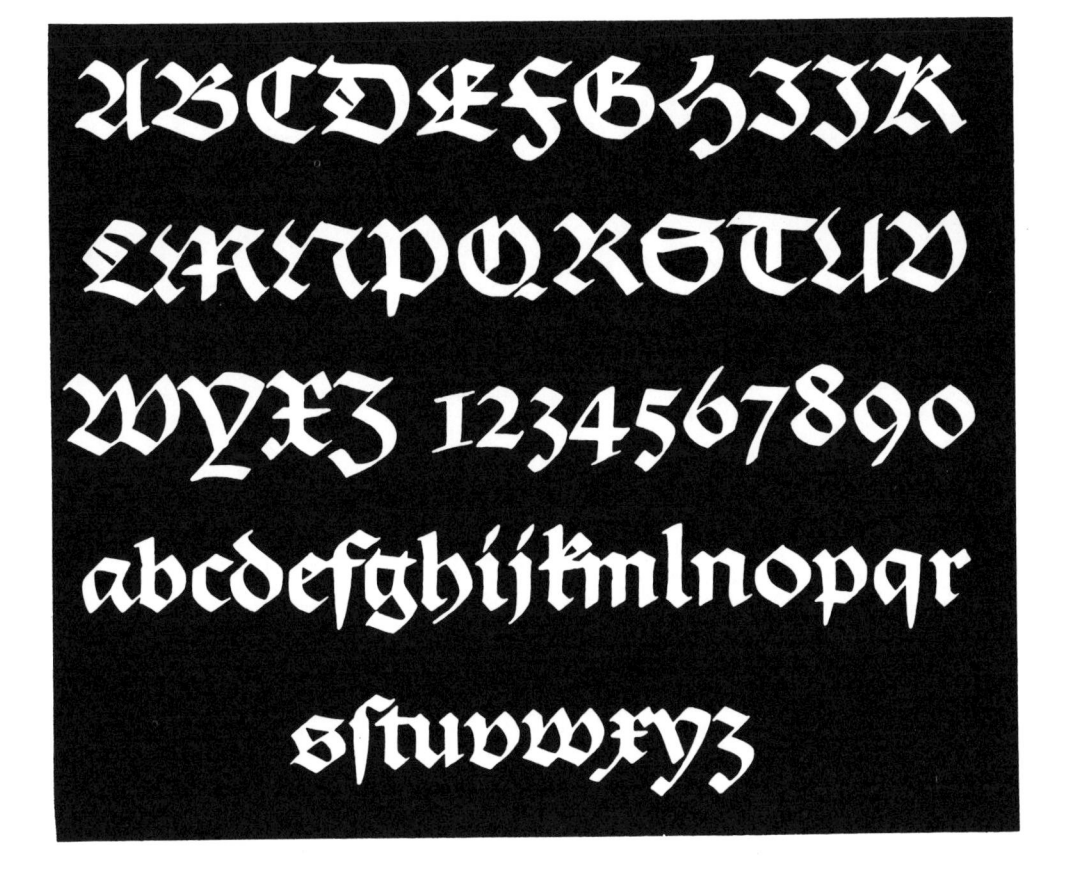

99

𝕾𝖈𝖙𝖊̄ 𝕸𝖎𝖍𝖆𝖊𝖑 𝖆𝖗𝖈𝖍𝖆𝖓𝖌𝖊
𝖋𝖊𝖓𝖉𝖊 𝖒𝖊 𝖎𝖓 𝖕̄𝖑𝖎𝖔:𝖛𝖙 𝖓𝖔𝖓

Figure 258

𝕮𝖑𝖆𝖗𝖎𝖘𝖘. 𝕻𝖎𝖈𝖙𝖔𝖗𝖎𝖘 𝖊𝖙 𝕲
𝕬𝖑𝖇𝖊𝖗𝖙𝖎 𝕯𝖚𝖗𝖊𝖗𝖎, 𝖉𝖊 𝖛𝖆

Figure 259

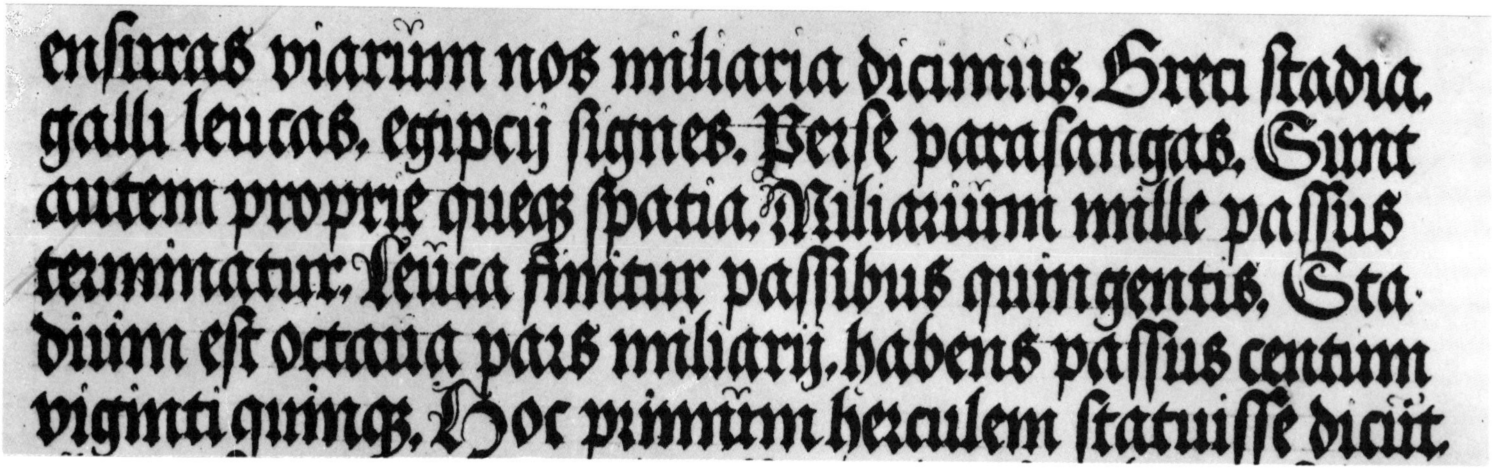

ensuras viarum nos miliaria dicimus. Grea stadia. galli leucas, egipcij signes. Perse parasangas. Sunt autem proprie queg spatia. Miliarium mille passius terminatur. leuca finitur passibus quingentis. Stadium est octaua pars miliarij. habens passus centum viginti quing. Hoc primum herculem statuisse dicit

Figure 260

𝕴𝖈𝖍 𝖜𝖎𝖑𝖑 𝖒𝖎𝖈𝖍 𝖍𝖎𝖓𝖋𝖚𝖗 𝖍𝖚𝖙𝖙𝖊𝖓 𝖜𝖔𝖑
𝕯𝖆𝖘 𝖎𝖈𝖍 𝖓𝖎𝖈𝖍𝖙 𝖑𝖊𝖎𝖈𝖍𝖙 𝖒𝖊𝖗 𝖐𝖔𝖒𝖊𝖓 𝖘𝖔𝖑
𝕴𝖓 𝖊𝖎𝖓 𝖘𝖈𝖍𝖎𝖋𝖋𝖑𝖊𝖎𝖓 𝖆𝖚𝖋 𝖉𝖆𝖘 𝖜𝖆𝖘𝖘𝖊𝖗

Figure 261

Figure 262

---

258 *Type from the prayerbook of Emperor Maximilian, 1515. From Albert Kapr,* Johann Neudörffer the Elder. *Leipzig, 1956.*

259 *Type from Dürer's* Proportionslehre, *1534. From Albert Kapr,* Johann Neudörffer the Elder.

260 *Fraktur, detail of "Clipanicana maior," from the sixteenth-century pattern book* Proba centum scripturarum *by Leonhard Wagner. Facsimile edition. Leipzig: Insel Verlag, 1963.*

261 *Type from the Theuerdank, 1517. From Albert Kapr,* Johann Neudörffer the Elder.

262 *Ductus.*

On page 101:

263 *Fraktur. Study by the author.*

264 *Fraktur. Study by the author.*

265 *Fraktur. Study by the author.*

266 *Upper Rhine style. Study by the author.*

267 *Upper Rhine style (the capitals do not match). Study by the author.*

268 *Variation of a fraktur. Study by the author.*

On pages 102 and 103:

269 *Renaissance fraktur initials by Paul Frank, Nuremberg, 1601. From Petzendorfer,* Schriften-Atlas *(Type atlas). Stuttgart: Verlag Julius Hoffmann, n.d.*

Figure 263 abcdefghijklmnopqurſstuvwxyz

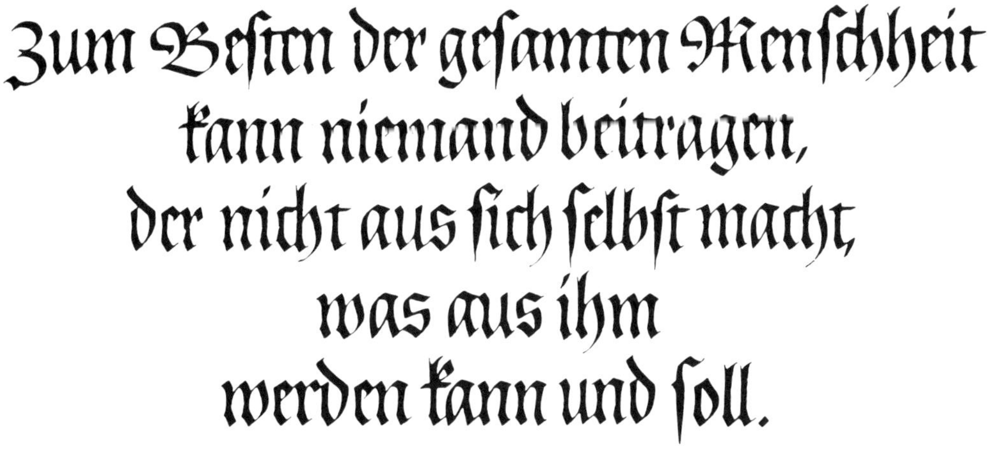

Zum Beſten der geſamten Menſchheit
kann niemand beitragen,
der nicht aus ſich ſelbſt macht,
was aus ihm
werden kann und ſoll.

Figure 264

Figure 265 Brandenburgiſche Konzerte

Figure 266 abcdefghijkmnlopqurſstuvwxyz

Figure 267 Alte Tänze

Figure 268 abcdefghijklmnopqrſstuvwxyz

Fraktur was used for the Book of Hours of Emperor Maximilian, 1513, by Schönsperger. Its forms were probably influenced by Leonhard Wagner and Vincenz Rockner (Figure 258). Also by Schönsperger is the type for the Theuerdank, 1517 (Figure 261). Neudörffer created the model for the types used in Dürer's *Proportionslehre* (1522 and 1534). The cutter was Hieronymus Andrea (Figure 259).

During the baroque era fraktur type moved further from its calligraphic roots and became increasingly flourished and elaborate.

Neoclassicism, under the influence of engravings on copperplates, favored forms that to us seem fragile, colorless, and artificial. Up to the nineteenth century fraktur type were the staple of German printers; today they are barely in use any more, displaced by classical roman letters. Among the typefaces still in use are the Alt Fraktur of Luther (1708), that of Breitkopf (eighteenth century), and Ernst Schneidler's Zentenar (1936).

When you draw these letters, pay special attention to the ligatures between the letters m, n, u, and h. Emphasize the differences between narrow lowercase and wide uppercase letters, and try to create a field of tension within capitals that contain large and small units.

101

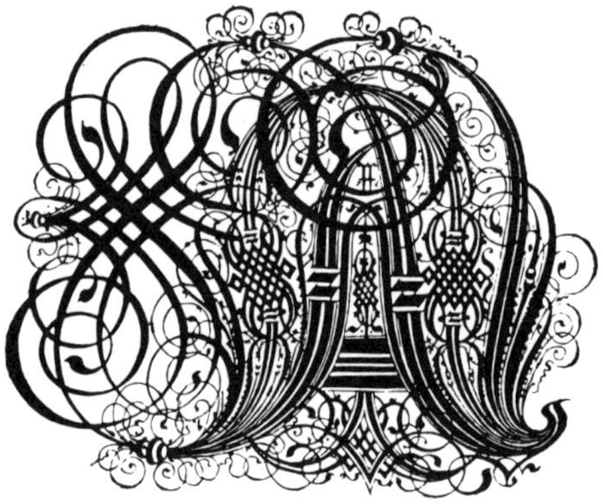
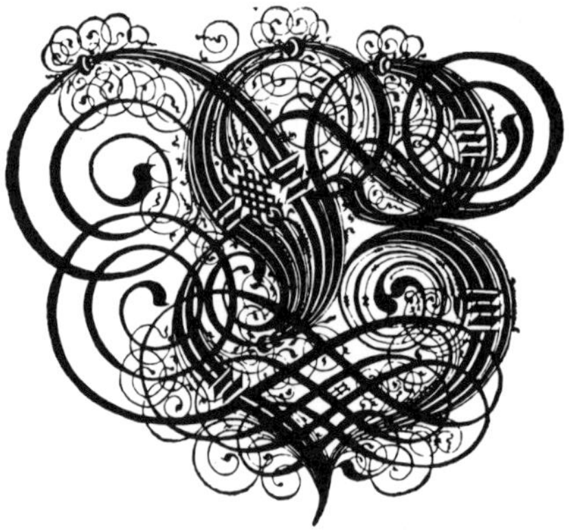
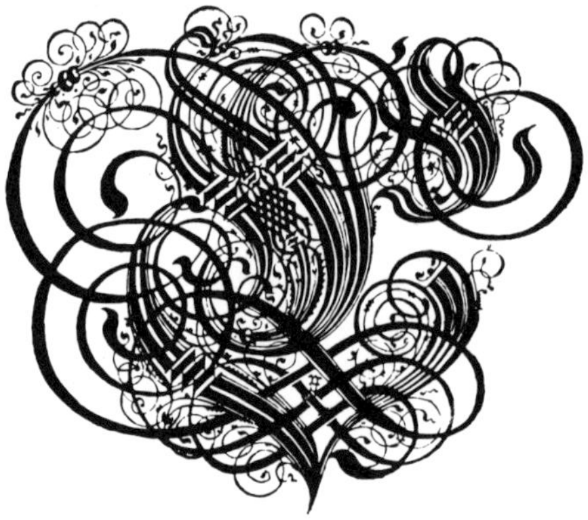
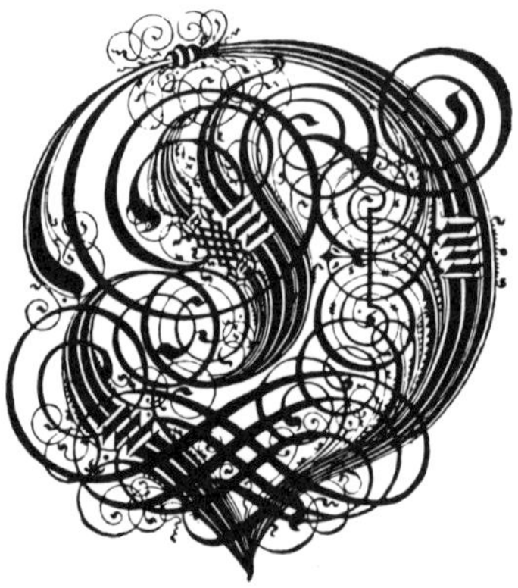
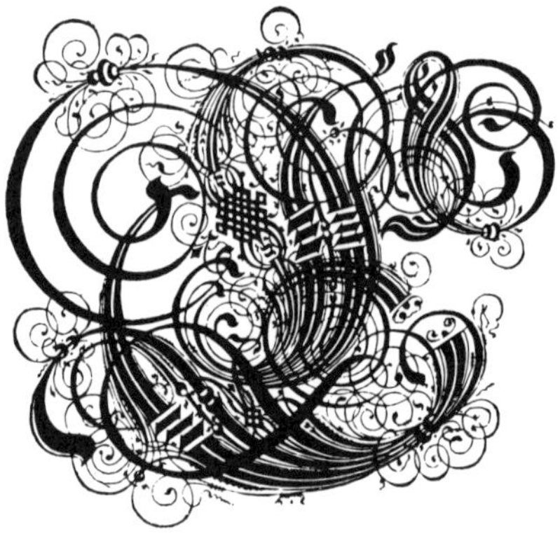
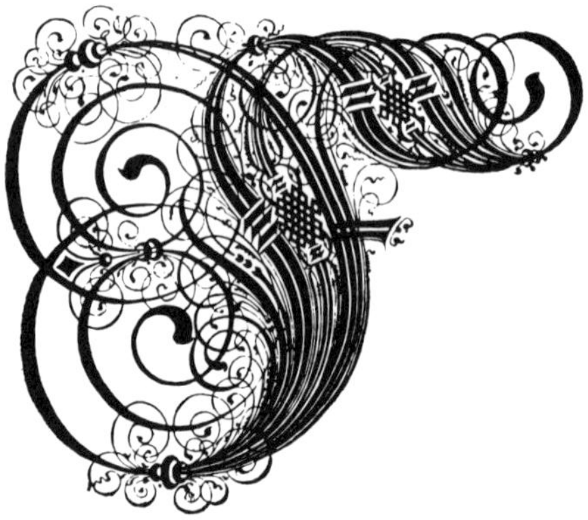
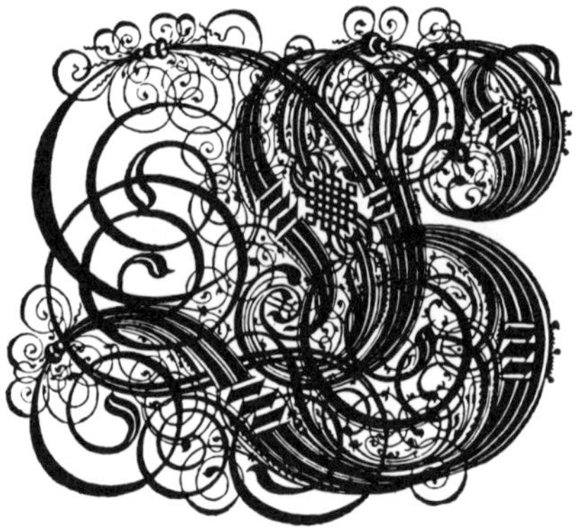
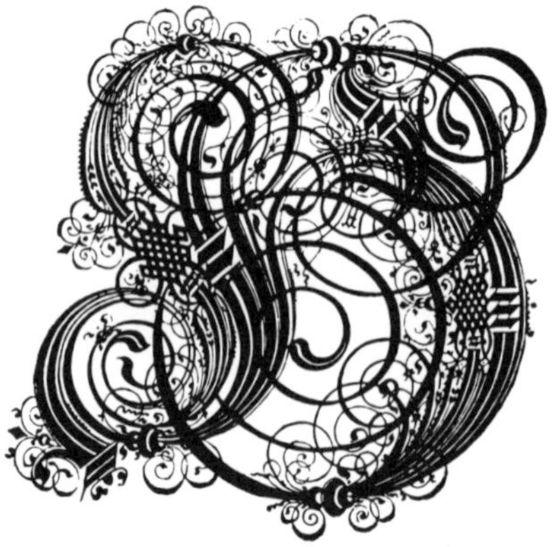
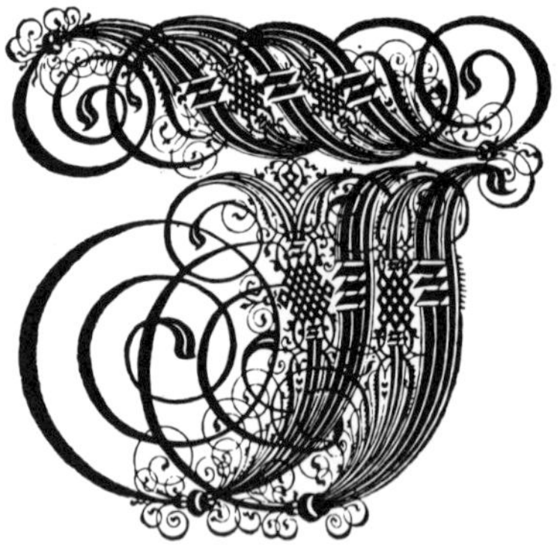
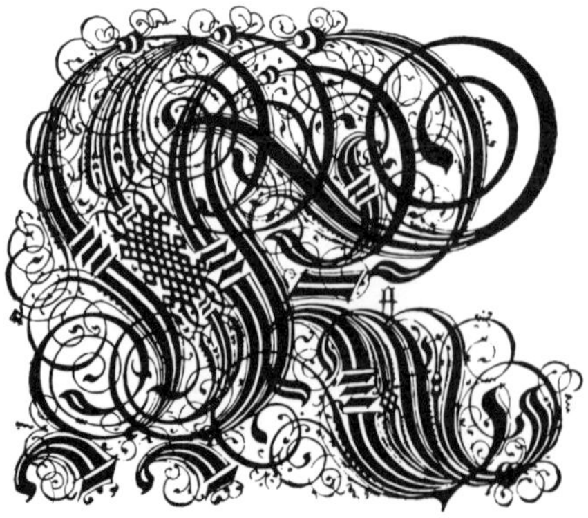
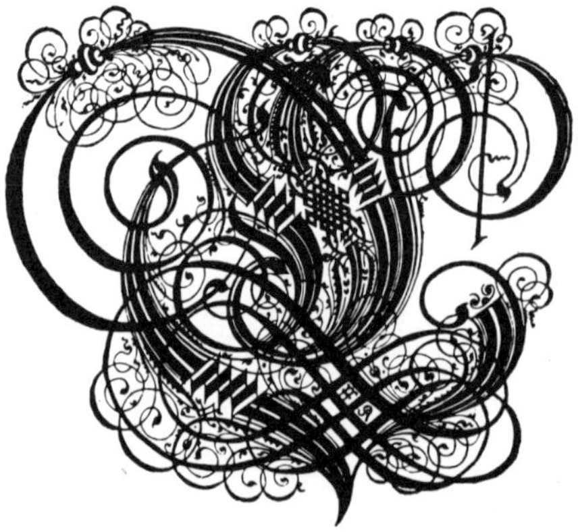
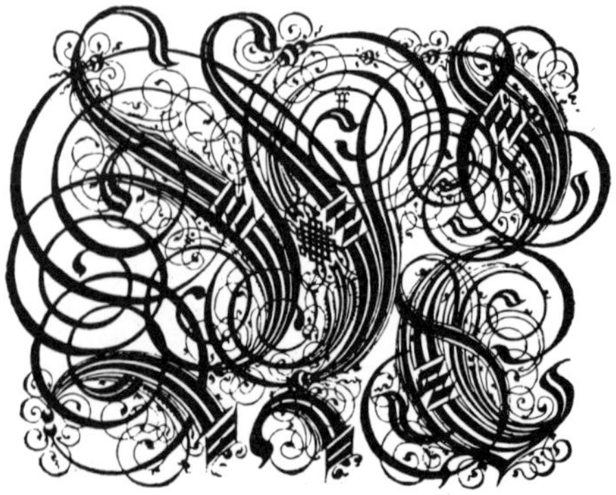

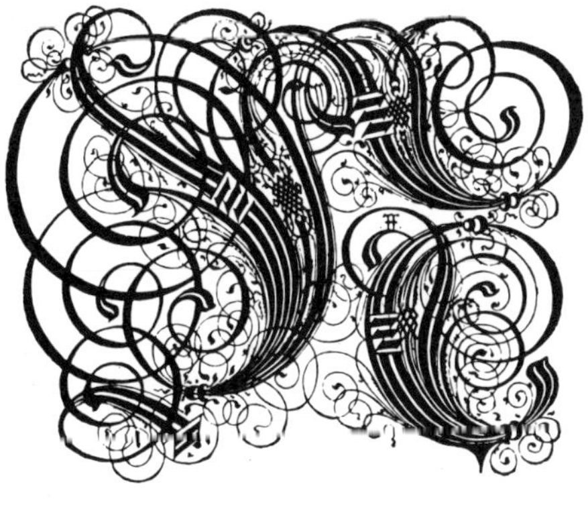
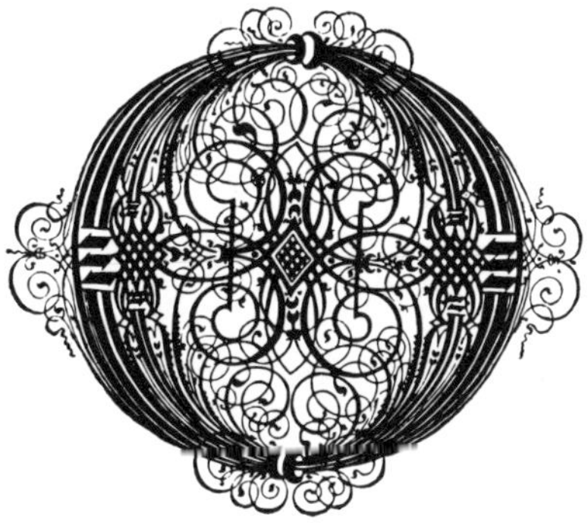
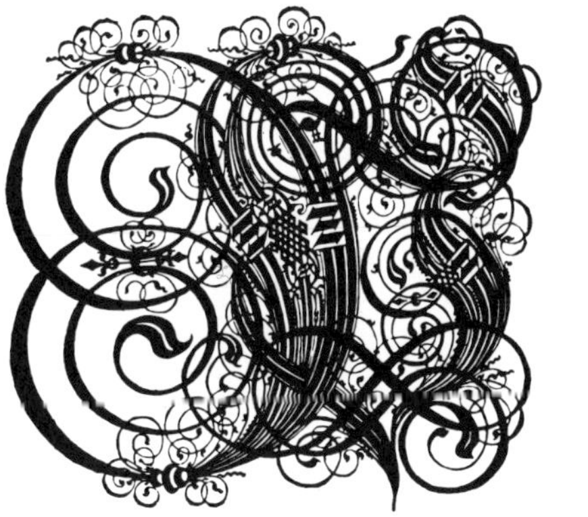
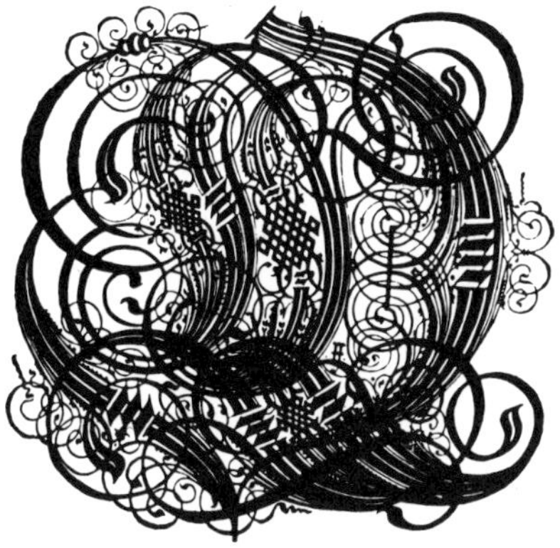
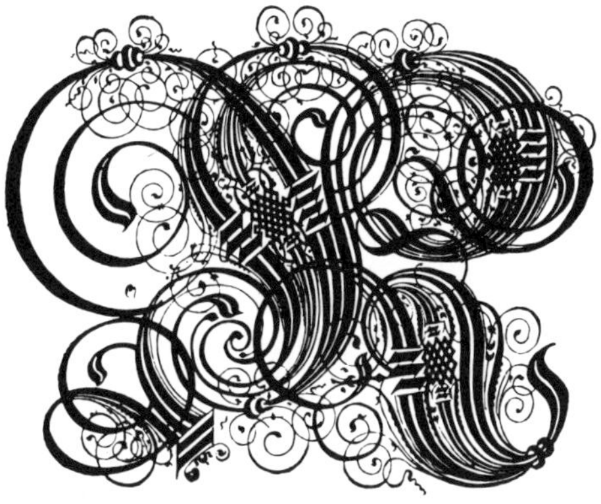
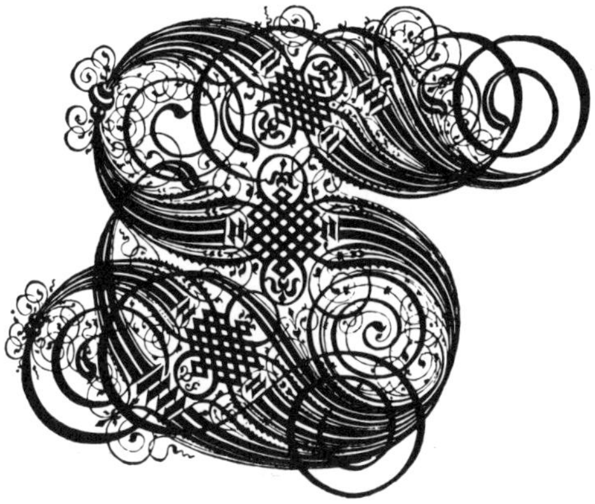
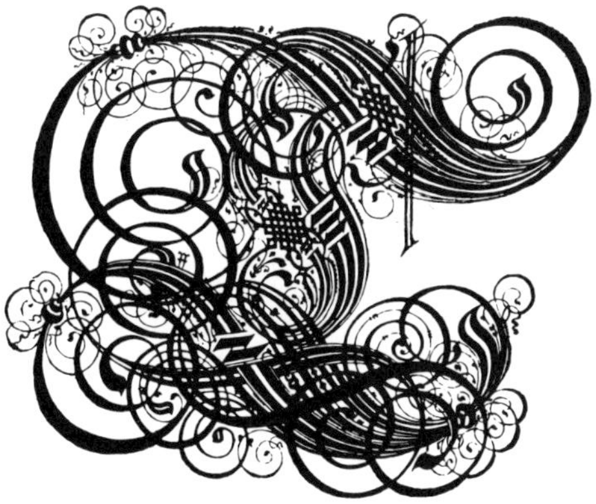
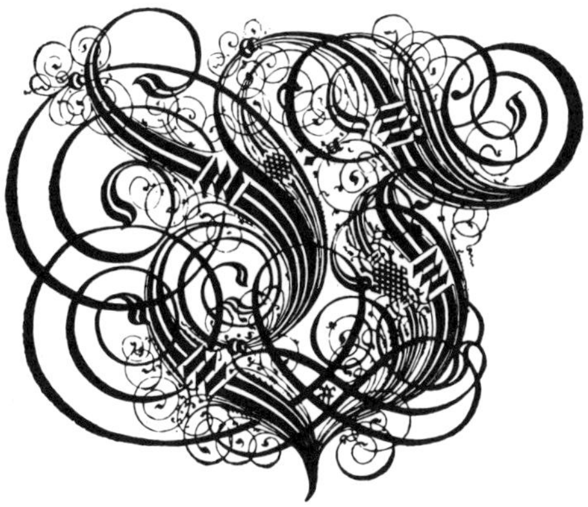
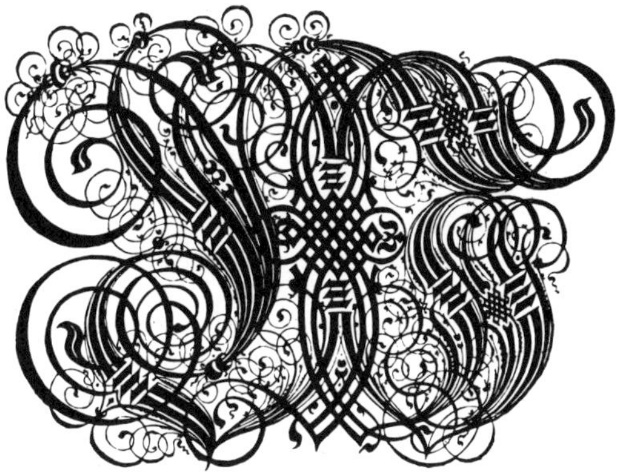
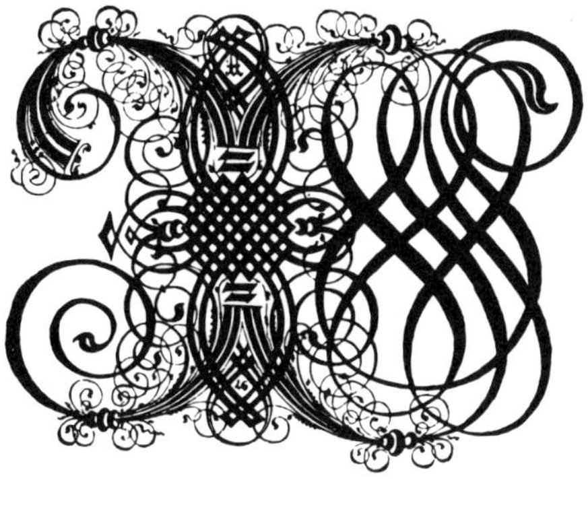
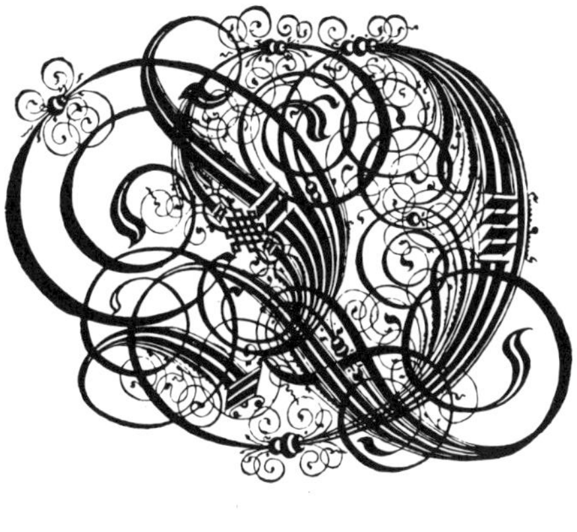
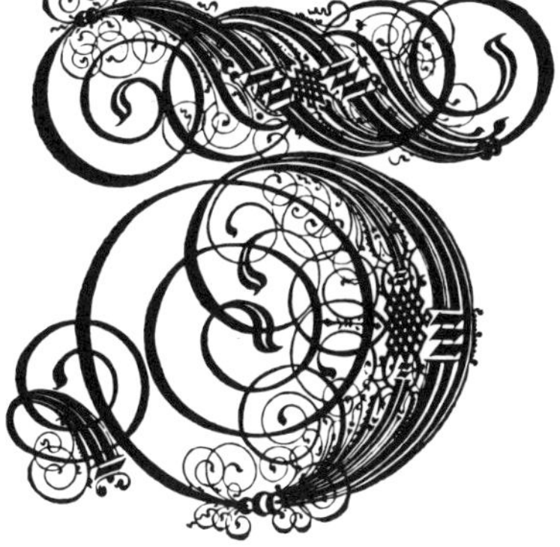

270 *Fraktur capitals. Study by the author.*

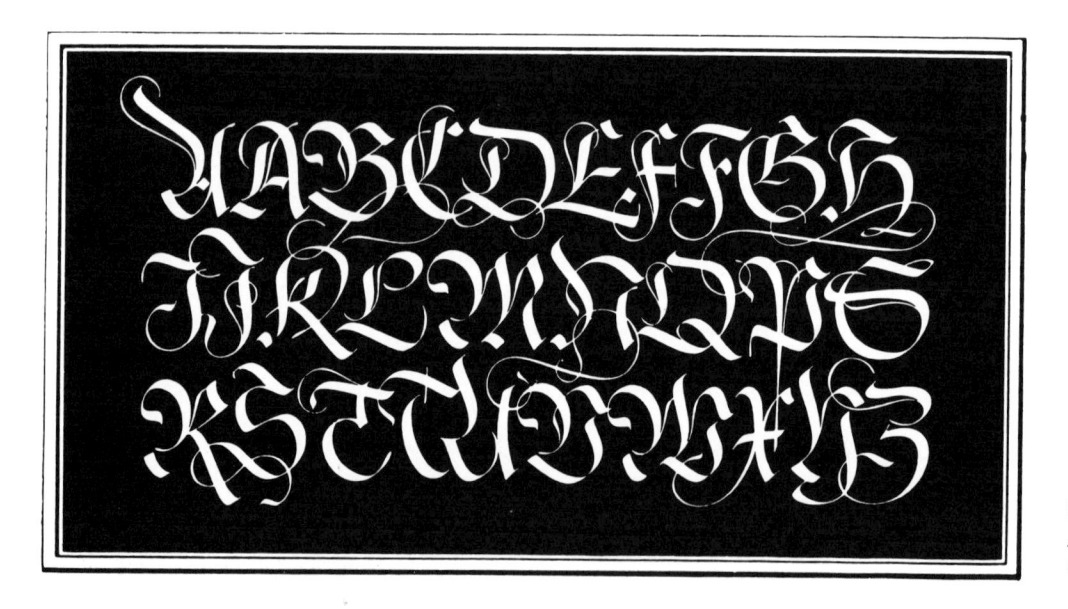

271 *Fraktur capitals (reduced) by Hermann Zapf. From* Pen and Graver. Alphabets & Pages of Calligraphy. *New York, 1952.*

272 *Alt Fraktur (36-point). D. Stempel AG, Frankfurt/M., Typesetting by Andersen Nexö, Leipzig.*

Chancery Cursive and German Kurrent

In Germany chancery cursive of the sixteenth century was a variation of fraktur, enriched by elements of Gothic script. The stems in x-height were either vertical or slanted to the left or to the right. Ligatures could be pointed or have a variety of arched forms. Many mixed types developed. The use of pointed quills and familiarity with engravings made the lettering of the seventeenth century ever more precise and delicate. A constant slant to the right was adopted, and x-heights shrank in favor of elongated ascenders and descenders. These tendencies intensified during the eighteenth century until the chancery cursive finally became identical with Kurrent, the common German handwriting.

Kurrent, in the sixteenth century similar to chancery cursive if somewhat more fluent, went through similar changes, until the steel nib, introduced in the nineteenth century, and the influence of English scripts turned it into the lettering style that was taught in German schools until the beginning of the twentieth century.

273 *Chancery script (historic form). From Wolfgang Fugger's* Schreibbüchlein, *facsimile edition, Leipzig, 1958.*

274 *Type page by Rudo Speman.*

RENAISSANCE ROMAN AND ITALIC

Renaissance Roman

As part of the pervasive reorientation that took place during the fifteenth century in Italy in practically all areas of learning and culture, the art of lettering was also affected. The Italian humanists found much of the ancient literature written in Carolingian minuscule and adopted this as book type (to paraphrase Muzika, *lettere antiche nuove*). The type was purged of all archaic remnants, the sequence of moves became conscious, letters like a, g, and r assumed a new shape. Capitals from the Roman monumental capitals were added, the stems of the minuscules were fitted with serif-like strokes. The Gothic tradition was overthrown in several stages. Arabic numerals first appeared in the fifteenth century.

During the sixteenth century the printed word supplanted the handwritten one in the production of books. Ever since, the history of letters is the history of printing type, which continues to draw inspiration from handlettering. Until the end of the nineteenth century, however, type forms moved ever further from their origins at the time of Gutenberg.

The first Venetian types show a strong effort to copy handlettering as closely as possible, for example, the Jenson roman of 1470. Characteristic of early Venetian forms are the slanted eye of the letter e, still evident in Centaur, a modern version of the Jenson. But the engraver of Aldus Manutius's type for *De Aetna* (1495) and *Hypnerotomachia Poliphili* (1499) started using his tools not only to re-create but to create. Bold type became pronounced, serifs more differentiated. Capitals were modeled after classical roman types.

The *Hypnerotomachia* served as model for the French Renaissance roman. The type created by and after Garamond (1544) soon could be found all over Europe.

In use today are replicas of the Renaissance roman types by Aldus Manutius, Bembo and Poliphilus. French types include replicas of Garamond and Jannon. Other modern creations using French and Venetian inspirations are Weiss Roman, Palatino, Trump Mediaeval, and, in Germany, Tschörtner.

Figure 275

Figure 276

Figure 277

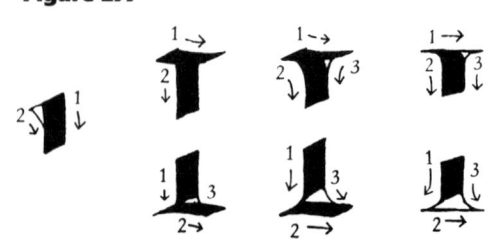

275 Humanist minuscule (historic form). From Arndt/Tangl, Schrifttalfeln zur Erlung der lateinischen Paläographie, Berlin, 1904.

276 Roman of Nicolas Jenson, Venice, 1470. From D.B. Updike, Printing Types, 3rd edition. London, 1962.

277 Ductus of crossbars and serifs.

278 Renaissance roman. Study by the author.

ABDEGHJKLMNOPRSTUVWXYZ

abcdefghijklmnopqrstuvwxyz

Über die Nachahmung. Der nur Nachahmende, der nichts zu sagen hat zu dem, was er da nachahmt, gleicht einem armen Schimpansen, der das Rauchen seines Bändigers nachahmt und dabei nicht raucht. Niemals nämlich wird die gedankenlose Nachahmung eine wirk-

abcdefghijklmnopqrstuvwxyzß

ABCDEGHIJKLMNOPRSTUVWXYZ

Figure 278

279 Uppercase alphabet (reduced) by Ludovico Vicentino (Arrighi), 1523.

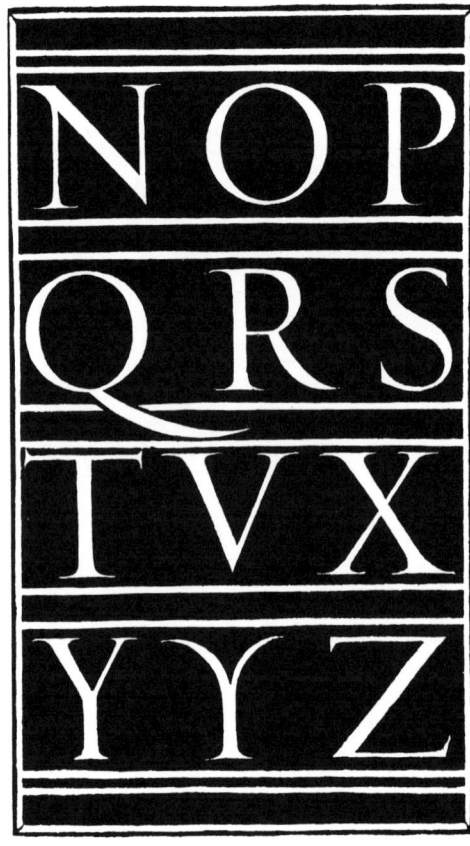

280 Roman lowercase letters from Il perfetto Scrittore of Giovanni Francesco Cresci, 1570.

ABCDEFGHIJKLMNOPQRSTUV
WXYZ   1234567890
abcdefghijklmnopqrstuvwxyz

ABCDEFGHIJKLMNOPQRSTU
VWXYZ   1234567890
abcdefghijklmnopqrstuvwxyz

ABCDEFGHIJKLMNOPQRSTUVWXYZ
1234567890
abcdefghijklmnopqrstuvwxyz

281 *Bembo (36-point). The Monotype Corporation. Typesetting by Andersen Nexö, Leipzig.*

282 *Garamond (36-point). Typoart, Dresden. Typesetting by Typoart, Dresden.*

283 *Chinese roman by Yu Bing-nan. Typoart Dresden/Institut für Buchgestaltung, Leipzig. Typesetting by Hochschule für Grafik und Buchkunst, Leipzig.*

ABCDEFGH
IJKLM
NOPQRSTU
VWXYZ
abcdefghijk
lmnop
qurstvwxyz
ß & .,!?
1234567890

284 Leipzig, designed by Albert Kapr, Typoart Dresden/Institut für Buchgestaltung, Leipzig.

285 Monument, roman capitals designed by Oldrich Menhart, Grafotechna.

On pages 112 and 113:

286 Humanist italic. From v. Larisch, Beispiele künstlerischer Schriften aus vergangenen Jahrhunderten (Artistic alphabets from the past), Vienna, 1910.

287 Renaissance italic from the Schreibbüchlein of Ludovico Vicentino (Arrighi), Rome, 1523.

288 Spanish modification of the italic by Francisco Lucas, Madrid, 1577. From Jan Tschichold, Treasury of Alphabets and Lettering. Reprint. Copyright © 1952, 1965 by Otto Maier Verlag.

289 Renaissance italic. Study by Harald Brödel.

ABCDEFGH
IJKLMNOPQR
STUVWXYZ
:{&}:
1234567890
BÁSNĚ
JANA
NERUDY
·
O.MENHART

Figure 285

Humanist Italic

Neocarolingian minuscules and Gothic scripts are equal contributors to the scripts of the Renaissance. The pages of the Italian humanists speak of strength, pride, and valor. Masters like Ludovico Vicentino (Arrighi), Giovanni Antonio Tagliente, Giambattista Palatino, and the Spaniard Francisco Lucas elevated the italic script under the name of cancellaresca, or chancery cursive, to a formal art.

The connected version, in which flourishes decorate ascenders, uses strictly vertical capitals. In unconnected styles capitals are slanted, ascenders are shorter and have serifs. Mixed forms exist, and highly decorated initials are popular. The Spanish style is somewhat softer than the Italian.

Cursive typefaces, called italic, have been produced since 1500. Francesco Griffo cut the first one for the Venetian printer and publisher Aldus Manutius. Between 1524 and 1526 Arrighi cut type that had a great influence on the details of other contemporary types. Both models are responsible for the French character of the Renaissance italics. In France Granjon, with his type of 1543, must be mentioned as eminent master of the italic.

An independent book type in its origins, created to save space, the italic sank to the level of a supplementary type towards the end of the sixteenth century. To this day it is used primarily for emphasis.

Figure 286

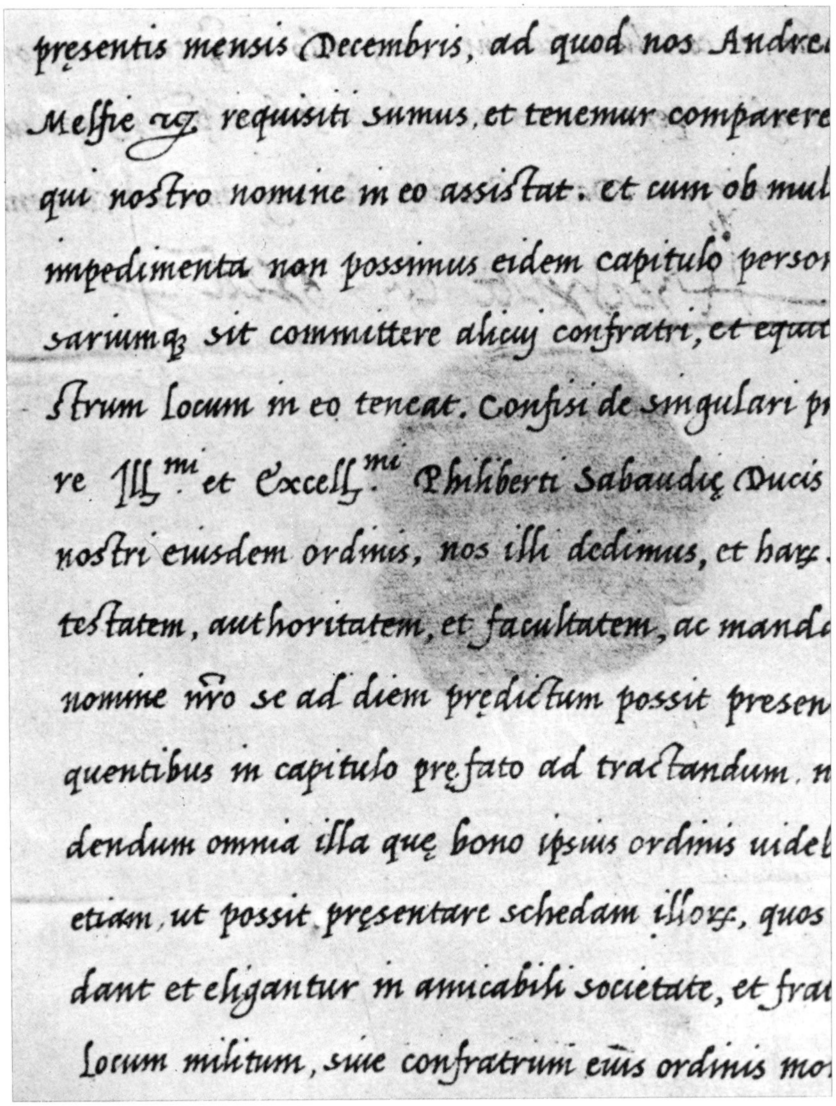

Figure 287

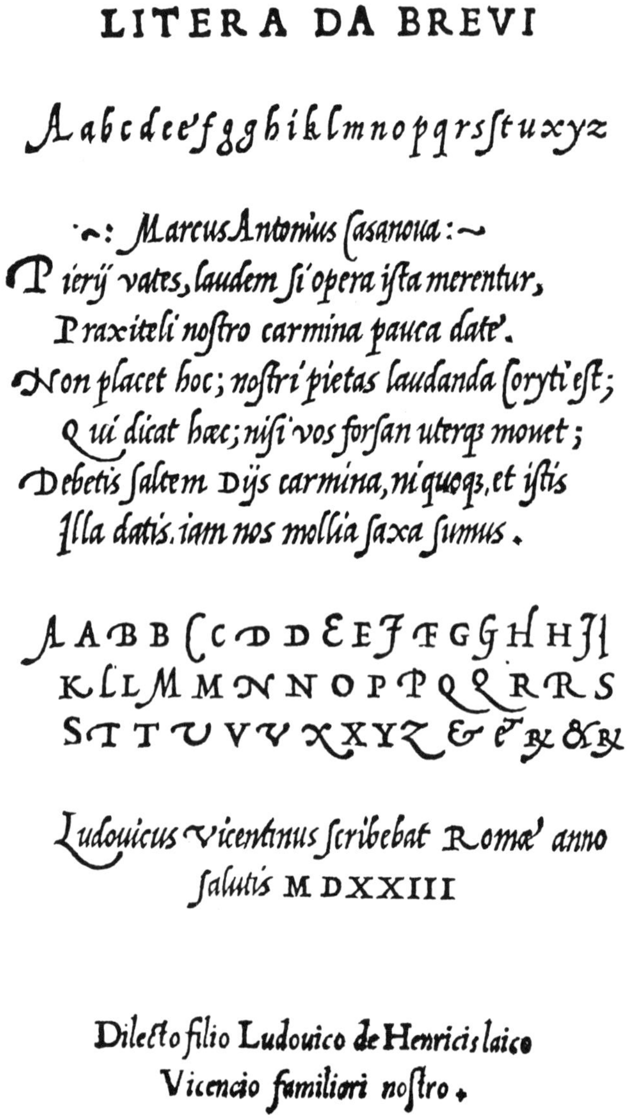

Figure 288

-: abc con sus principios
A · co a · l k b · rc rod · ec
l s f r o g o o p p g l s h h
c c s s i i l l l i r n r m
r n c c o l l l p p y c o q u
r y s · ss l t i v u x
v v y y · z z z y
Fran.co Lucas me escre-
uia en madrid año 570

-: Bastarda grande llana :-
Obsecrote domina sancta
Maria mater Dei pietate
plenissima, summi regis fi-
lia, mater gloriosissima, m.a-
ter orphanorum, consola-
tio desolatorum, via erran-
tiuz
Fran.co Lucas lo escreuia en
Madrid año de MD1XX

Figure 289

abcdefghijklmnopqurstuvwxyz

113

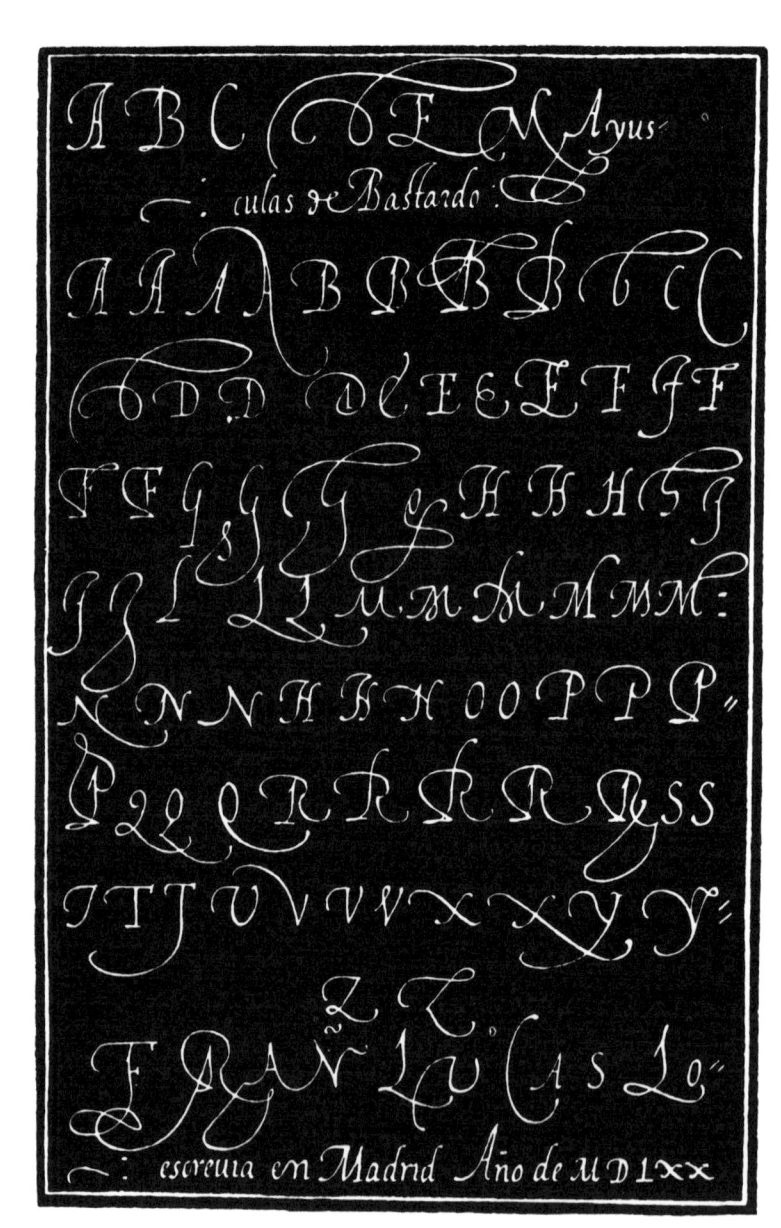

Figure 290

Figure 291

Figure 292

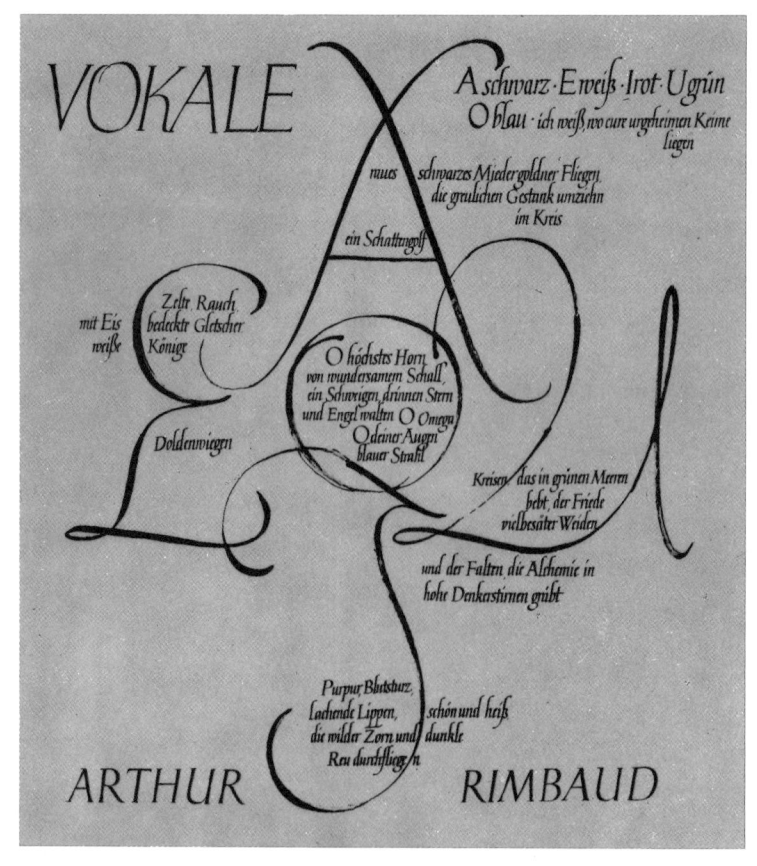

290 *Spanish italic modification. Capitals by Francisco Lucas, Madrid, 1577. From Jan Tschichold,* Treasury of Alphabets and Lettering. *Reprint. Copyright © 1952, 1965 by Otto Maier Verlag.*

291 *Renaissance italic with elements of Gothic cursive. Calligraphy by Albert Kapr.*

292 *Renaissance italic by Irmgard Horlbeck-Kappler.*

293 *Italic. Study by the author.*

ABCDEFGHIJK
LMNOPQR
TSUVWXYZ
DEZ

abcdefghijklmnopqu
rstuvwxyzßgvw

294 *Italic. Study by the author.*

Und die ungenaueste Wolke, das alltäglichste Wort,
das geringste Ding - Zwinge sie die Flügel zu schlagen.
Mache sie deinem Herzen ähnlich.

Man muß geradezu verschwenderisch sein in der
Lust am Lernen. Sparsamkeit in dieser Richtung
und Geiz sind das Schlimmste, was einem Künstler
geschehen kann Denn bald wird er aufhören, wenn

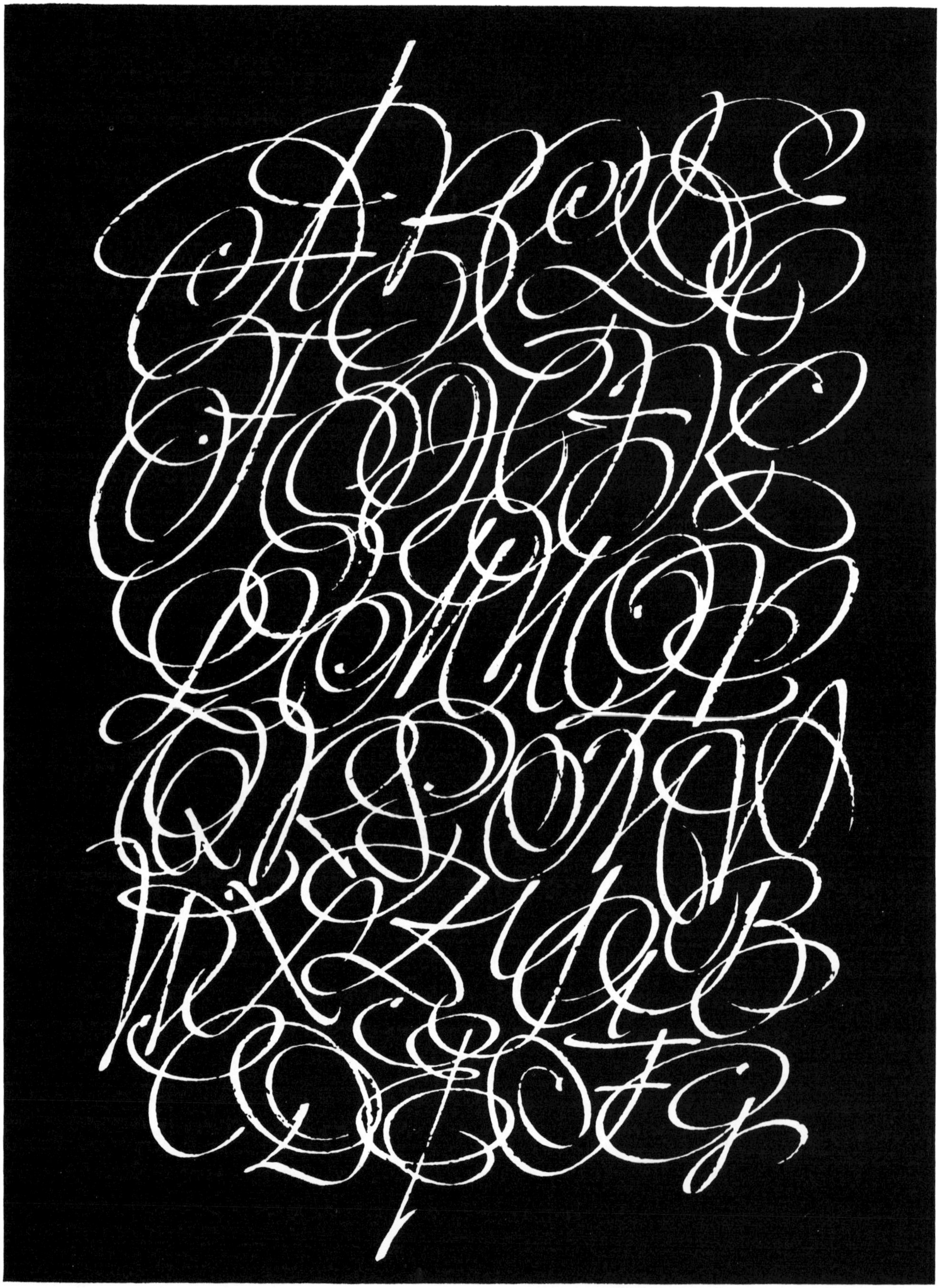

ABCDEFG
HIJKLMN
OPQRSTU
VWXYZ

abcdefghijklmno
pqrstuvwxyzäöü
chckffififlflaßtttz
1234567890

295 *Italic capitals. Study by Villu Toots.*

296 *Stentor (60-point), typeface based on italic drawn with a brush. Designed by Heinz Schumann. Typoart Dresden. Typesetting by Typoart Dresden.*

ABCDEFGHIJKLMNOP
QRSTUVWXYZ
abcdefghijklmnopqrstuvwxyz
1234567890

297 *Bembo italic (36-point). The Monotype Corporation. Typesetting by Andersen Nexö, Leipzig.*

ABCDEFGHIJKLMNO
PQRSTUVWXYZ
abcdefghijklmnopqrstuvwxyz
1234567890

298 *Garamond italic (28-point). Typoart Dresden. Typesetting by Typoart Dresden.*

ABCDEFGHIJKLMOPQRST
UVWXYZ
abcdefghijklmnopqrstuv
wxyz   1234567890

299 *Faust italic (28-point). Designed by Albert Kapr. Typoart Dresden/Institut für Buchgestaltung, Leipzig. Typesetting by Hochschule für Grafik und Buchkunst, Leipzig.*

FROM THE BAROQUE TO THE NINETEENTH CENTURY

Baroque Roman

The form canon of the baroque roman stands between that of the Renaissance and the neoclassical roman. There is little resemblance to commonly known baroque style elements. The face of the Renaissance roman underwent important changes during the seventeenth century. There were practical, which is to say technical, reasons, one of them the new use of the graver. The typefaces of the Hungarian Nicholas Kis (creator of the face erroneously called Janson), and the Dutchmen van Dyck and Fleischmann are darker and less elegant than their French predecessors, but they are very legible and usable. Ascenders and descenders are reduced in favor of a larger x-height, the thick/thin contrast is heightened, round serifs are straightened, and the points exaggerated. The stress of the round forms is steeper too. English modifications of this type were created by Caslon and Baskerville at the beginning of the eighteenth century, French ones at the end of the eighteenth century by Grandjean for the Imprimerie Royale and by Fournier. Grandjean's Romain du Roi has importance as a transition to the neoclassical type. Among modern typefaces Times and Imprimatur should be mentioned.

300 *Baskerville roman and italic (36-point). D. Stempel AG Frankfurt/Main. Typesetting by Andersen Nexö, Leipzig.*

ABCDEFGHIJKLMNOPQRSTUV
WXYZ 1234567890
abcdefghijklmnopqrstuvwxyz

ABCDEFGHIJKLMNOPQRSTUVW
XYZ 1234567890
abcdefghijklmnopqrstuvwxyz

Neoclassical Roman

The aesthetic and rational spirit of neoclassicism is embodied in the types of Didot in France, Bodoni in Italy, and Walbaum in Germany. They impress us with precision and strength, but they are less lively than those of the Renaissance or baroque roman which the neoclassicists stripped of any traces of the graver. Characteristics include stark contrasts between thick and thin strokes, capitals of almost identical width, and horizontal hairlines for serifs even on lowercase letters.

An improved printing technique and smooth paper were the prerequisites for this development. Modern modifications of the neoclassical roman are among others: Amati, Corvinus, Figura, Pergamon, and Tiemann. Attempts were made to use width proportions of the Renaissance roman for an improved neoclassical type: Diotima and Athenaeum. The work is executed with an almost horizontal pen. A similar attempt is shown in the baroque roman of Figure 302.

301 *Didot roman capitals.*

302 *Roman alphabet in the baroque style. Study by Harald Brödel.*

303 *Neoclassical capitals. Calligraphy by Irmgard Horlbeck-Kappler.*

On pages 122 and 123:

304 *Bodoni roman and italic (36-point). Typesetting by Andersen Nexö, Leipzig.*

305 *Walbaum roman and italic (36-point). Typesetting by Andersen Nexö, Leipzig.*

306 *Eighteenth-century decorative roman. Caslon, Fry & Steele, Stephenson Blake & Co., Sheffield.*

Figure 301

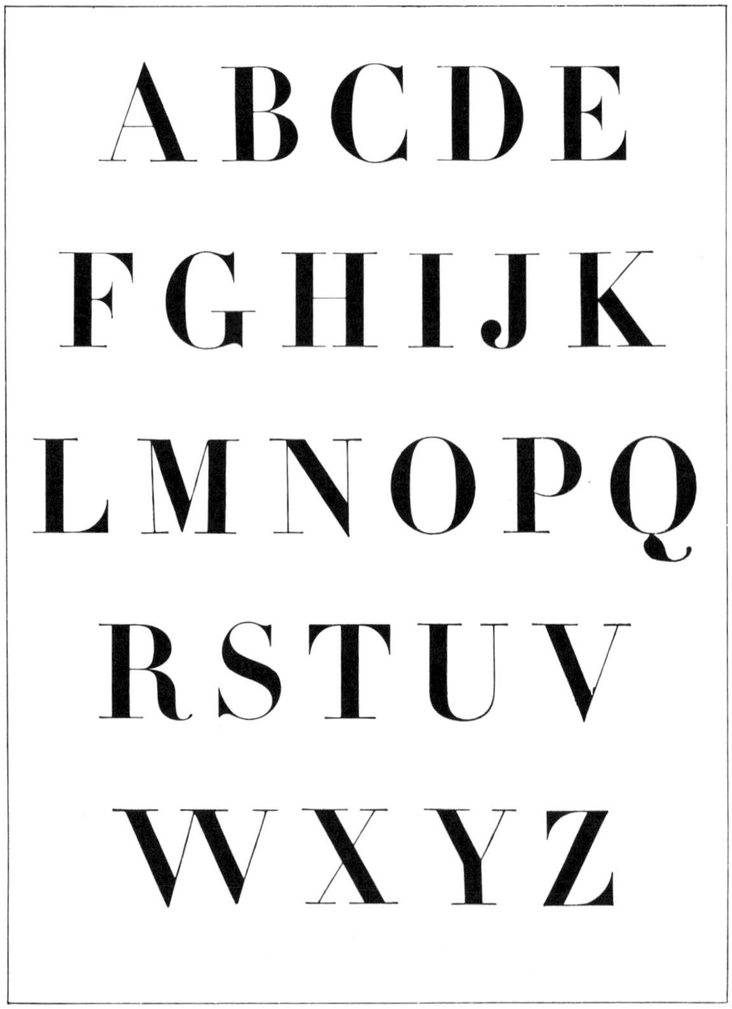

Figure 302

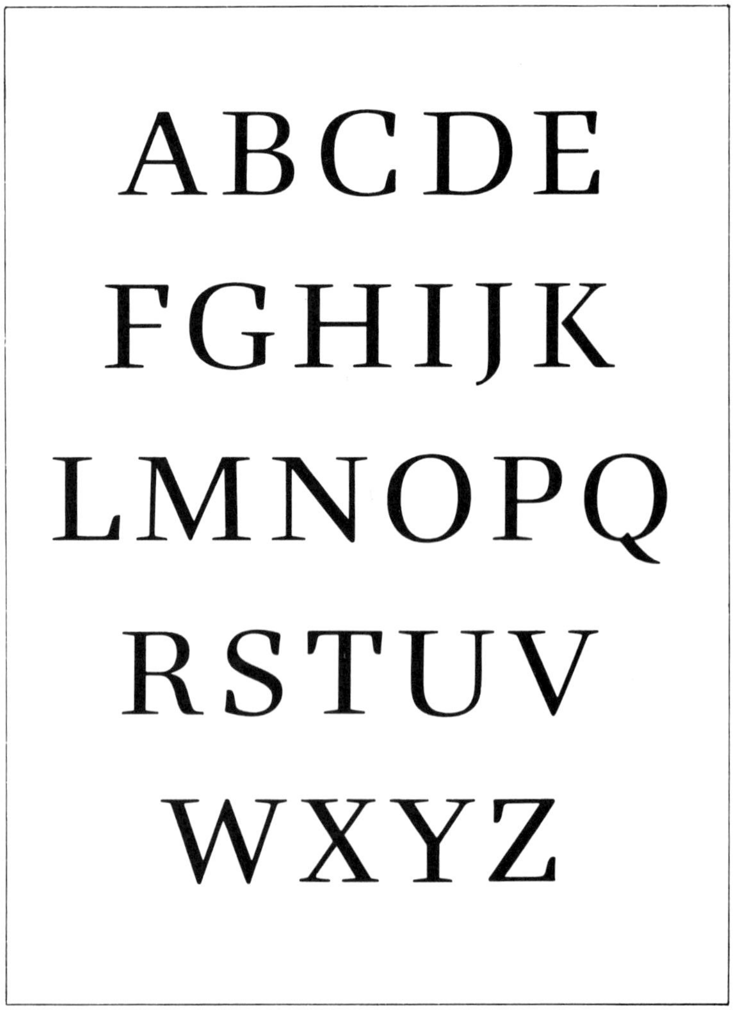

ICH SCHWOERE BEI APOLLON · DEM ARZT · UND BEI ASKLEPIOS · BEI HYGIEIA UND
PANAKEIA UND BEI ALLEN GOETTERN UND GOETTINNEN
DIE ICH ZU ZEUGEN ANRUFE · DASS ICH NACH BESTEM VERMOEGEN UND URTEIL
DIESEN EID UND DIESE VERPFLICHTUNG ERFUELLEN WERDE

ICH WERDE DEN · DER MICH DIESE KUNST LEHRTE · MEINEN ELTERN GLEICHACHTEN
MIT IHM DEN LEBENSUNTERHALT TEILEN UND IHN · WENN ER NOT LEIDET
MITVERSORGEN · SEINE NACHKOMMEN MEINEN EIGENEN BRUEDERN GLEICHSTELLEN
UND SIE DIE HEILKUNST LEHREN · WENN SIE DIESE ERLERNEN WOLLEN · OHNE ENTGELT
UND OHNE VERTRAG · RATSCHLAG UND VORLESUNG UND ALLE UEBRIGE BELEHRUNG
WILL ICH AN MEINE EIGENEN SOEHNE UND AN DIE MEINES LEHRERS
WEITERGEBEN · SONST ABER NUR AN SOLCHE SCHUELER · DIE NACH AERZTLICHEM
BRAUCH DURCH DEN VERTRAG GEBUNDEN UND DURCH DEN EID
VERPFLICHTET SIND

MEINE VERORDNUNGEN WERDE ICH TREFFEN ZU NUTZ UND FROMMEN
DER KRANKEN · NACH BESTEM VERMOEGEN UND URTEIL UND
VON IHNEN SCHAEDIGUNG UND UNRECHT FERNHALTEN · ICH
WERDE NIEMANDEM · AUCH NICHT AUF SEINE BITTE HIN · EIN
TOEDLICHES GIFT VERABREICHEN ODER AUCH NUR
EINEN SOLCHEN RAT ERTEILEN · AUCH WERDE ICH NIE
EINER FRAU EIN MITTEL ZUR VERNICHTUNG
KEIMENDEN LEBENS GEBEN ·
HEILIG UND REIN WILL ICH MEIN LEBEN
UND MEINE KUNST BEWAHREN · WELCHE
HAEUSER ICH BETRETEN WERDE · IMMER WILL
ICH EINTRETEN ZUM HEILE DER KRANKEN
MICH ENTHALTEN JEDER VORSAETZLICHEN
UND VERDERBLICHEN SCHAEDIGUNG · AUCH
ALLER WERKE DER WOLLUST AN DEN LEIBERN VON
FRAUEN UND MAENNERN · FREIEN UND SKLAVEN ·
WAS ICH BEI DER BEHANDLUNG ODER AUCH AUSSERHALB
DER BEHANDLUNG IM LEBEN DER MENSCHEN SEHE ODER
HOERE · WERDE ICH · SOWEIT MAN ES NICHT AUSPLAUDERN DARF
VERSCHWEIGEN UND SOLCHES ALS GEHEIMNIS BETRACHTEN

WENN ICH DIESEN EID ERFUELLE UND NICHT VERLETZE · MOEGE MIR
IM LEBEN UND IN DER KUNST ERFOLG ZUTEIL WERDEN UND
RUHM BEI ALLEN MENSCHEN FUER EWIGE ZEITEN
WENN ICH IHN ABER UEBERTRETE UND MEINEIDIG WERDE
SO GESCHEHE MIR DAS GEGENTEIL

ABCDEFGHIJKLMNOPQRSTUVW
XYZ   1234567890
abcdefghijklmnopqrstuvwxyz

*ABCDEFGHIJKLMNOPQRSTUVW
XYZ   1234567890
abcdefghijklmnopqrstuvwxyz*

Figure 304

Figure 305

ABCDEFGHIJKLMNOPQRSTUV
WXYZ   1234567890
abcdefghijklmnopqrstuvwxyz

*ABCDEFGHIJKLMNOPQRST
UVWXYZ   1234567890
abcdefghijklmnopqrstuvwxyz*

ABCDE

GHIJKL

MNOPQ

RSTU

VWXYZ

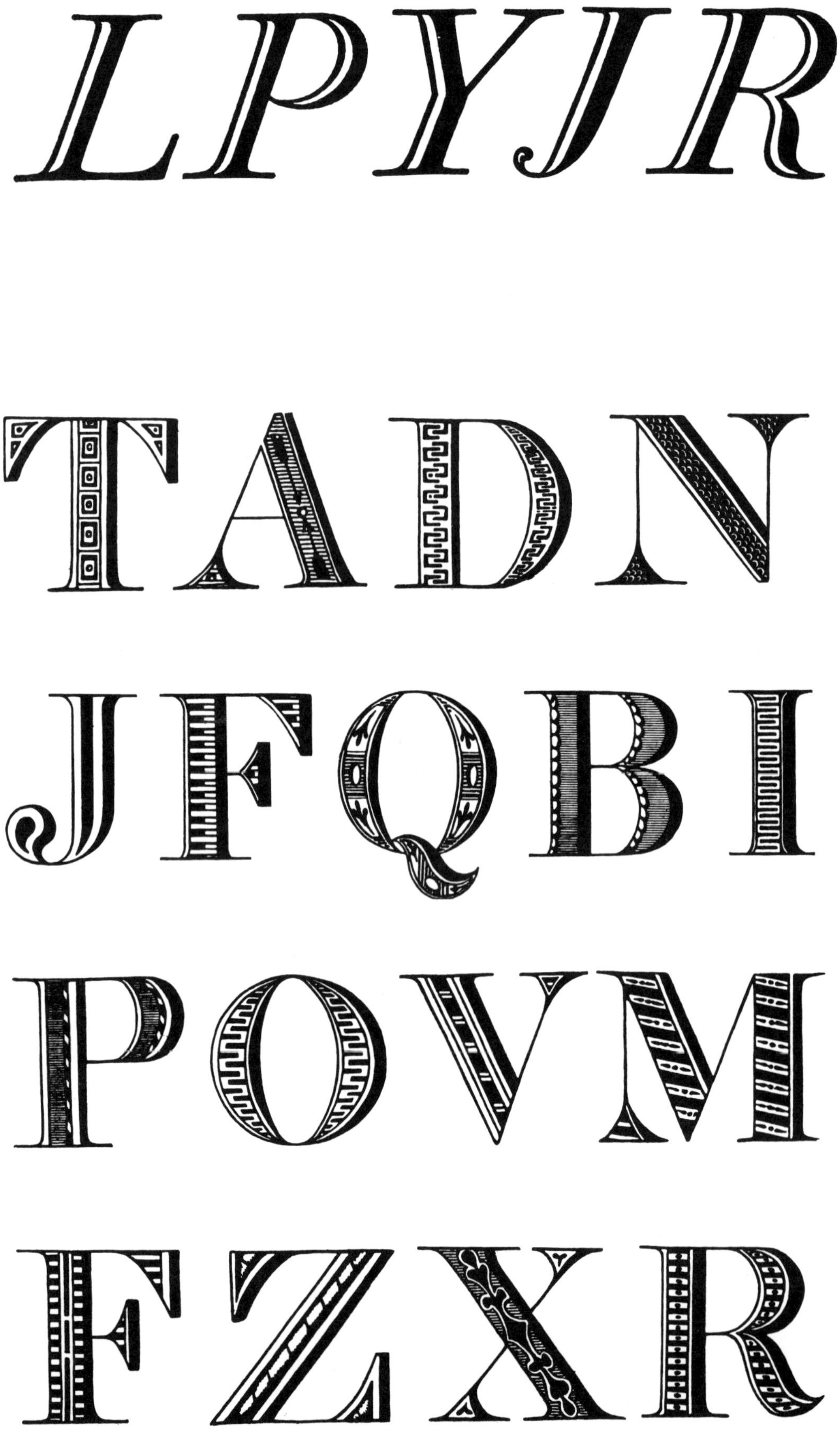

LPYJR

TADN

JFQBI

POVM

FZXR

Figure 307

Seventeenth- to Nineteenth-Century Scripts

During the time of the high baroque, scripts were not written with wide pens, but with pointed ones. Ascenders and descenders were embellished decoratively and with great virtuosity, while the middle sections of the letters slowly deteriorated. Outstanding calligraphers of the time were the Dutchman Jan van den Velde, the Frenchmen Francois Desmoulins and Louis Barbedor, and the Spaniard Juan de Polanco. Engraving techniques tempted them into increasingly perfected artistry.

The Latin cursive lost its vitality during the eighteenth and nineteenth centuries. The pointed quill, and later the steel nib, produced sharp lines and ever more similar forms. Engraving became the desired model of handwriting. England led the way in this development and drew all of Europe behind her. The effects are still felt in spite of several reforms of the methods used to teach schoolchildren. Several typefaces were modeled after the English pattern.

307 Top: Decorative italic capitals. P.S. Fournier, Paris, 1766. Bottom: Decorative roman capitals, about 1800. From Alphabets for Signwriters, Artists and Illuminators, London, 1956.

308 Top: Calligraphic handwriting. Copperplate engraving from van den Velde, Haarlem, 1649. Center: Handwriting by Bourgoin, about 1810. Bottom: Calligraphic handwriting. Copperplate engraving from de la Chambre, Haarlem, 1649.

Figure 308

125

309 Decorative uppercase alphabet by Andrade, Lisbon, 1722.

Andrade

310 *English roundhand. Study by Harald Brödel, 1969.*

NORST
VWXYZ
GU&(.,;:!?-»«')
67890

qurstuvwxyzäöüßtr

Nineteenth-Century Display Styles

One of the side effects of the industrial revolution in England was the advertising industry. Advertisements, posters, and other printed matter created a demand for eye-catching and effective typefaces, some of them larger than ever before. European firms were quick to adopt and further develop the new English typefaces. First came the bold roman of neoclassical style, followed by Egyptian, Italian, and Tuscan faces in many variations. Their names allow inferences to the original purpose. Egyptian, one assumes, was created to satisfy the interest in all things Egyptian that arose in England after Napoleon's ill-fated campaign, Kapr says. Type production became the domain of industry. Foundries were engaged in competition with each other, which was beneficial in the beginning, but catastrophic in terms of artistic quality soon afterwards. Catalogs hawked well-designed and questionable alphabets side by side; type design degenerated into mere technique.

Display typefaces are of great importance again today. The following section presents an overview of the most useful display types. Basic forms can easily be varied and adapted to specific applications.

Bold Roman Neoclassical

An extreme contrast between thick and thin distinguishes bold roman neoclassical types, also called fat faces, from the neoclassical roman. It can be quite elegant, nevertheless. Keep the interior spaces narrower than the stroke width. Variations are shown in Figures 314 through 325.

Egyptian and Italienne

The first of these types were probably created before 1806 in the foundry of Robert Thorne (according to Muzika). The thick/thin contrast is reduced to a minimum. The stress of the round forms is vertical, the capitals are matched to each other in width. The serifs are rectangular in shape (slab serifs) and the same width as the crossbars. Earlier versions have straight serifs, later ones are bracketed. A common name for these was Clarendon, after the type of that name. Both basic forms in various weights and excellent new copies have been produced during the last few years by several foundries: Volta by Bauer, Neutra by Typoart, Clarendon by Haas, Egizio by Nebiolo, as well as some faces for typewriters. The illustrations show that Egyptian is a good starting point for a variety of decorative variations. Only some later Italienne types are worth mentioning. They differ from Egyptian in the horizontals, which are bolder than the verticals.

Tuscan

In 1815 Tuscan type was introduced in England by Vincent Figgins. Its peculiarity is bifurcated stems and serifs, a device already popular during antiquity, the Renaissance, and the baroque era. Tuscan faces can be developed from Egyptian or from Italienne.

Sans Serif

There are different opinions about the first cut of sans serif, or grotesque, type. Albert Kapr quotes the year 1803 in his *Klassifizierung der Satzschriften*[4] and the company as Thorne, while Frantisek Muzika credits the Caslon foundry in the year 1834 as the creator of the first sans serif alphabets, which were cut as "Egyptian without serifs" (Muzika). (The first known example of a sans serif was from William Caslon IV's 1816 specimen book.) The horizontals are only slightly narrower than the verticals, the stress is vertical, the width of the capitals is almost equal. Many variations are possible through changes in stroke width and letter width. The condensed medium sans serif is the most useful of them.

Historic sans serif typefaces are graced with a certain liveliness and charm, and it should not surprise us that type designers try to approach the historical wellspring of type yet again with creations such as Folio, Helvetica, and Univers, after they exploited all the possibilities of constructivism during the twenties and thirties. Univers is of special interest, because it contains vertical and diagonal bars that are differentiated in accordance with the principle of changing stroke width in the roman, a characteristic that has influenced the way we all see type. The width of the capitals is less uniform than in other sans serif styles.

---

4. Albert Kapr, "Die Klassifikation der Druckschriften" (The classification of printing types), in *Schriftmusterkartei*. Leipzig: VEB Fachbuchverlag, 1967.

ABCDEFG
HIJKLMN
OPQRSTU
VWXYZ
abcdefghij
klmnopqrs
tuvwxyz
1234567890

311 *Extended bold neoclassical roman. Thorowgood, Stephenson Blake & Co.*

A A B C D E F
G H J K L M Y
M N O P Q R S
T U V V W W
a b c d e f g h i j k
l m n o p q r s t u v
w x y z

312 *Extended bold neoclassical italic. Thorowgood Italic, Stephenson Blake & Co., Sheffield.*

ABCDE
FGHIJKL
MNOPQRST
UVWXYZ
12345

313 *Condensed semibold neoclassical roman. American Type Founders.*

ABCDEFG
abcdeffifflg

314 Decorative variations of an extended bold neoclassical roman. Genzsch & Heyse, Hamburg, 1834.

ABCD

315 Decorative capitals of an extended bold neoclassical roman. F.A. Brockhaus, Leipzig, 1846.

AMDEB

316 Decorative capitals of an extended bold neoclassical roman. John T. White, New York, 1845.

BURG

317 Decorative capitals of an extended neoclassical roman. Joh. Enschedé en Zonen, Haarlem, 1841.

SCHELLINKHOUT

318 Extremely extended bold neoclassical roman capitals. Joh. Enschedé en Zonen, Haarlem, about 1841.

LECTUR

319 Decorative capitals of an extremely extended Egyptian, France, nineteenth century.

SLYONA

320 Decorative capitals of an extended bold neoclassical roman. R. Thorne, London, 1810.

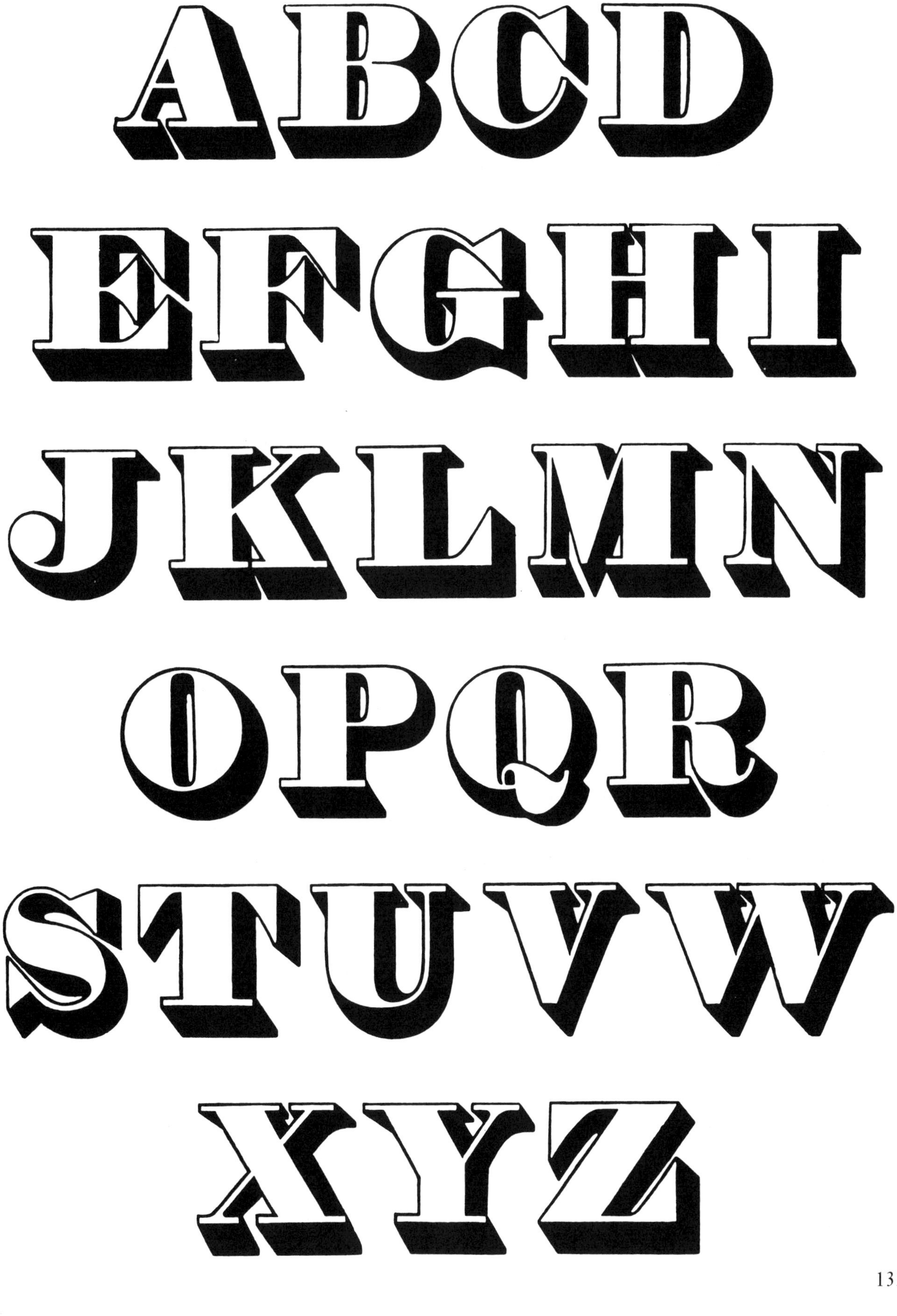

ABCD
EFGHI
JKLMN
OPQR
STUVW
XYZ

Figure 320

135

321 Decorative capitals of an extended bold neoclassical roman. F.A. Brockhaus, Leipzig, 1846.

322 Decorative capitals of a condensed bold neoclassical roman. F.A. Brockhaus, Leipzig, 1846.

323 Decorative capitals of an extended bold neoclassical roman. John T. White, New York, 1845.

324 Decorative capitals of an extended bold neoclassical roman. John T. White, New York, 1845.

325 Decorative capitals of an extended bold neoclassical roman. J. Gillé, Paris, 1828. Deberny & Peignot.

ABCDEFG
HIJKLMN
OPQRSTU
VWXYZ12
3456789

Figure 326

ABCDEFG
HIJKLMN
OPQRSTU
VWXYZÇ&

abcdefghij
klmnopqrs
tuvwxyzch
ßckäöü

1234567890

326 Neutra, an Egyptian face designed by Albert Kapr. Typoart Dresden.

327 Condensed semibold Egyptian. J.G. Schelter & Giesecke, Leipzig, 1897.

On pages 140 and 141:

328 Extended bold Egyptian decorative capitals (reduced). "Lettres ombrées ornées." J. Gillé, Paris, about 1810. Deberny & Peignot, Paris. From Jan Tschichold, Treasury of Alphabets and Lettering. Reprint. Copyright © 1952, 1965 by Otto Maier Verlag.

329 Extended bold Egyptian-Tuscan decorative capitals (mixed form). "Romantics." Paris, about 1830. Fonderie Typographique Français, Paris, XV. From Jan Tschichold, Treasury of Alphabets and Lettering. Reprint. Copyright © 1952, 1965 by Otto Maier Verlag.

ABCDEFG
HIJKLMN
OPQRSTU
VWXYZ
1234567890

Figure 327

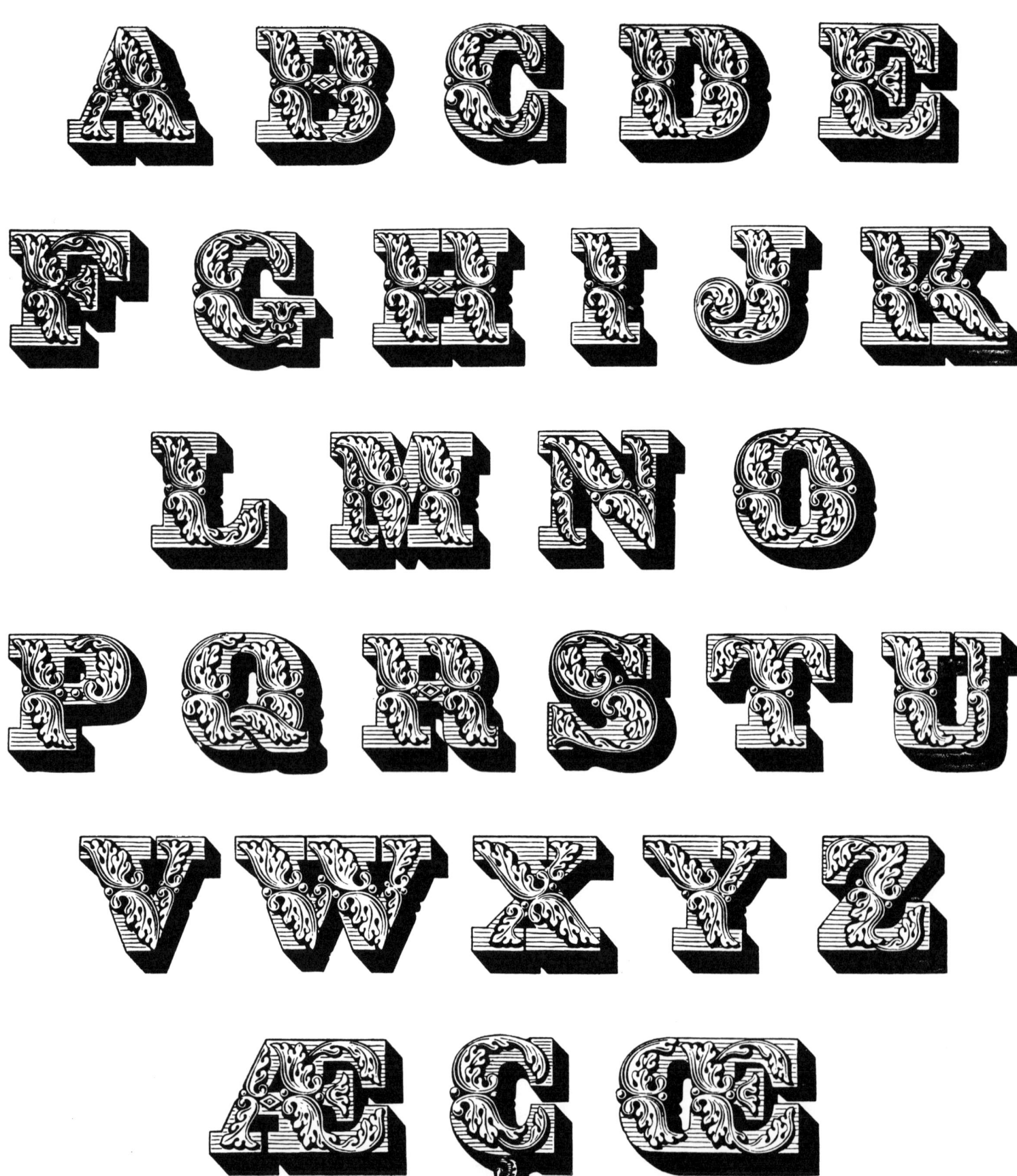

Figure 328

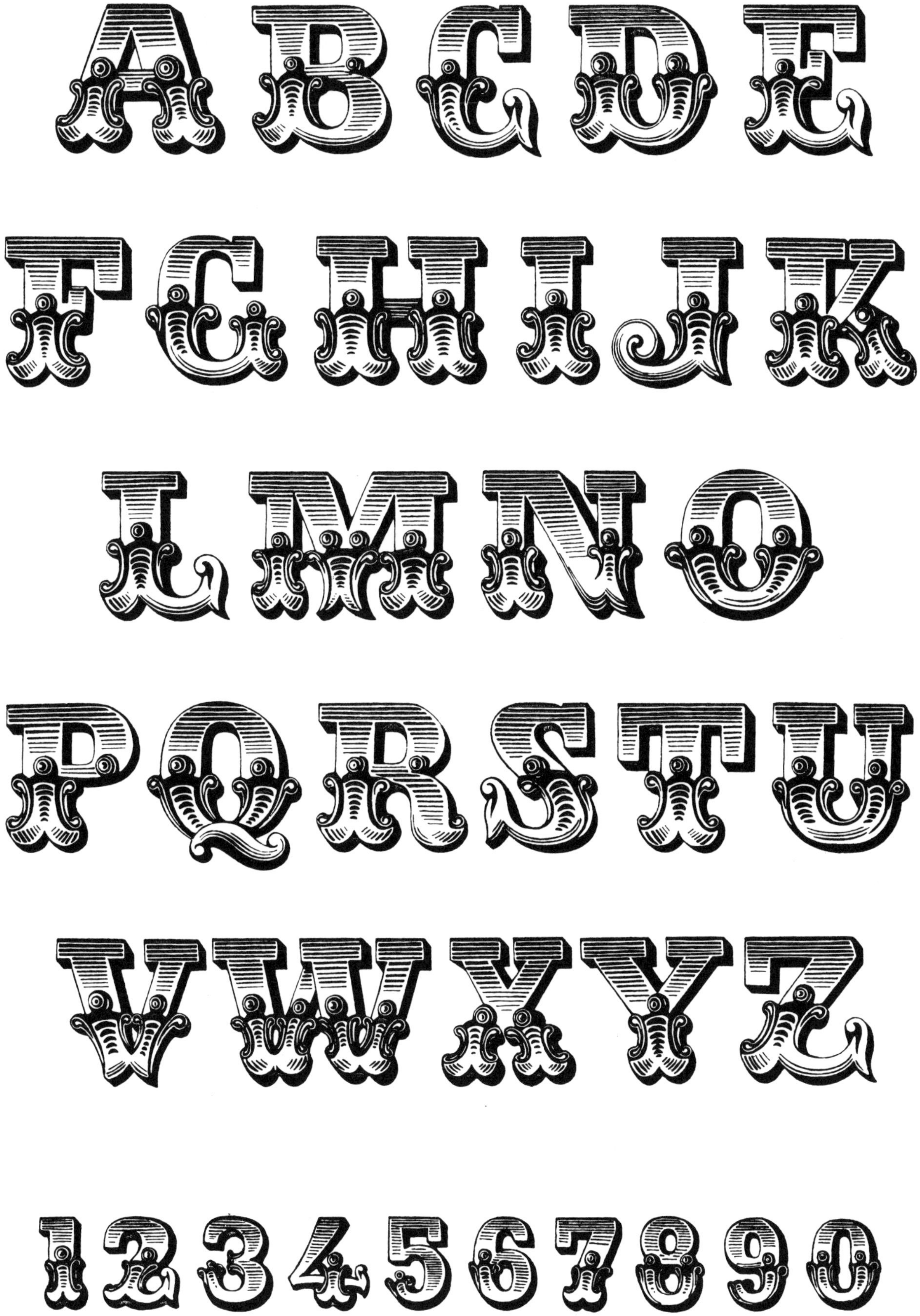

Figure 329

ABCDEF
GHIJKL
MNOPQR
STUVW
XYZÄÖÜ

330 *Decorative capitals of an extended bold Egyptian. F.A. Brockhaus, Leipzig, 1846.*

BARSINGEHO

331 *Decorative capitals of a bold Egyptian. Joh. Enschedé en Zonen. Haarlem, middle of the nineteenth century.*

ZEIST

332 *Decorative capitals of a bold Egyptian. Joh. Enschedé en Zonen. Haarlem, middle of the nineteenth century.*

333 *Bold Egyptian. Joh. Enschedé en Zonen, Haarlem, middle of the nineteenth century.*

APLEXMONDRCHT

334 *Decorative capitals of an extended bold Egyptian. France, nineteenth century.*

THIERY

335 *Decorative italic capitals of an extended bold Egyptian. John T. White, New York, 1845.*

AMERIC

336 *Decorative capitals of an extended bold Egyptian. F.A. Brockhaus, Leipzig, 1846.*

HAMBURG

WEIMT

337 *Decorative capitals of an extended bold Egyptian. Joh. Enschedé en Zonen. Haarlem, middle of the nineteenth century.*

GENMUIDJSP

338 *Decorative capitals of an extended bold Egyptian. John T. White, New York, nineteenth century.*

WYMING

339 *Decorative capitals of an extended bold Egyptian. Joh. Enschedé en Zonen. Haarlem, middle of the nineteenth century.*

HAARLEM

340 Tuscan. Genzsch & Heyse, Hamburg, 1834.

341 Decorative Tuscan. France, nineteenth century.

342 Decorative Tuscan. France, nineteenth century.

343 Decorative Tuscan. John T. White, New York, nineteenth century.

344 Tuscan. Decorative numerals. From Alphabets for Signwriters, Artists and Illuminators, London, 1956.

ABCDEFGHIJK
LMNOPQ
RSTUVWXYZ
abcdefghijklmn
opqrstuvwxyz
123456789

345 *Condensed Italienne, nineteenth century.*

ABCDEFGHI
KLMNOPQU
RSTVWXYZ
1234567890

346 *Alphabet in Egyptian-style italic. Designed by Volker Küster, 1966.*

abcdefghijk
lmnopqrstu
vwxyzßäöü
&.,:;-'()!?»«-*J

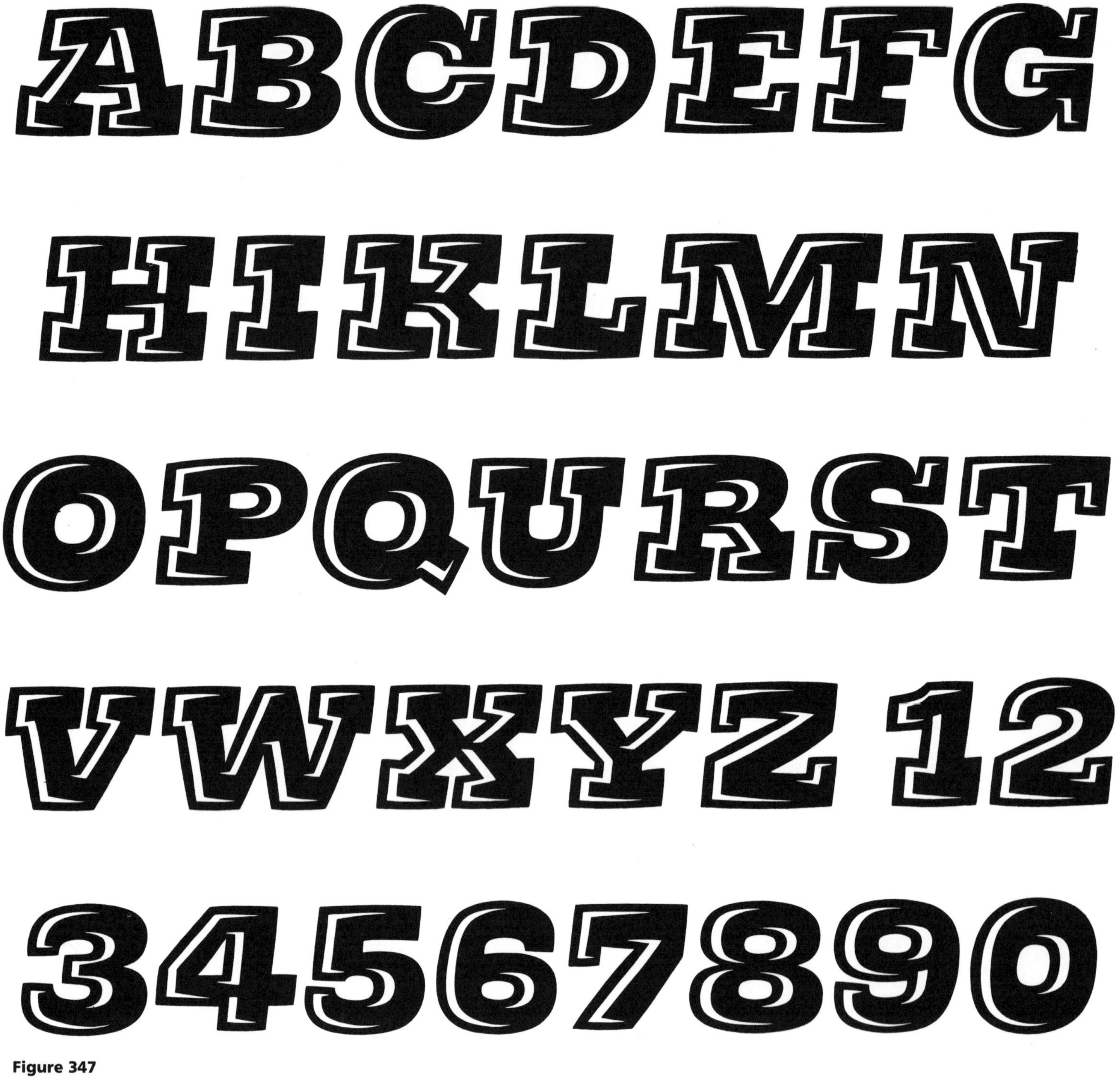

Figure 347

Figure 348

148

ABCDEFG
HIJKLMNOP
QRST
UVWXYZ
1234567890

abcdefg
hijklmnopqrst
uvwxyz&äöü

347 Decorative capitals. Variation of the alphabet in Figure 346. Designed by Volker Küster, 1966.

348 Decorative capitals. Another variation of the alphabet in Figure 346. Designed by Volker Küster, 1966.

349 Nineteenth-century sans serif. Haas, Munchenstein. From Jan Tschichold, Treasury of Alphabets and Lettering. Reprint. Copyright ©1952, 1965 by Otto Maier Verlag, Ravensburg.

Figure 350

ABC
DEFGHIJKL
MNOPQRST
UVWXYZ

abcdefghijkl
mnopqrsftu
vwxyz

1234567890

350 *Eckmann (art nouveau). Klingspor. Offenbach/Main, about 1900.*

351 *Futura (enlarged), by Paul Renner. Bauer, Frankfurt/Main, 1932.*

On pages 152 and 153:

352 *Akzidenz-Grotesk. H. Berthold AG, Berlin-Stuttgart. Originally cut about 1898. Reproduced by "diatype," 1961–67.*

ABCDEFG
HIJKLMNOP
QRSTU
VWXYZ
1234567890
abcdefg
hijklmnopqrst
uvwxyz

Figure 351

ABCDEFGHIJKLMNOPQRSTUVWXŽ
YZÄÖÜÈÕŠ1234567890»«„;:–!?"§&()[]
abcdefghijklmnopqrstuvwxyzß äöü ãéž

ABCDEFGHIJKLMNOPQRSTUVW
XYZÄÖÜØÅÇ1234567890(;:!?'–§&]
abcdefghijklmnopqrstuvwxyzßäöüåç

ABCDEFGHIJKLMNOPQRSTUVW
XYZÄÖÜĞË1234567890›‹;!?"§&)[
abcdefghijklmnopqrstuvwxyzß äöü

ABCDEFGHIJKLMNOPQRSTUV
WXYZÄÖÜ1234567890,;:!?äöü
abcdefghijklmnopqrstuvwxyzßí

ABCDEFGHIJKLMNOP
QRSTUVWXYZÄÖÜÅÉŠ
1234567890%;!?§&([›/
abcdefghijklmnopqrst
uvwxyzßäöüåéçš

АБВГДЕЖЗИЙКЛМНОПРСТУФХ
ЦЧШЩЭЮЯЁЪЫЬабвгдежзийкл
мнопрстуфхцчшщэюяёъыь

АБВГДЕЖЗИЙКЛМНОПРСТУФХ
ЦЧШЩЭЮЯЁЪЫЬабвгдежзийк
лмнопрстуфхцчшщэюяёъыь

ABCDEFGH
JKLMNOP
QRSTUVW
XYZJ ÄÖÜ
123
4567890

abcdefghijkl
mnopqrstuv
wxyzßäöü
&&%*†
(.:!?,;-–»„"/'´˘)

353 *Sans serif. Designed by Harald Brödel, 1968.*

ABCDEFGH
JKLMNOP
QRSTUVW
XYZJÄÖÜ
123
4567890

abcdefghijkl
mnopqrstuv
wxyzßäöü
&&%*†
(.:!?;-—„"»/´˘)

354 Sans serif. Designed by Harald Brödel, 1968.

ABCDEFG
HIJKLMNO
PQRSTUV
WXYZJ
ÄÖÜ123
4567890

abcdefghij
klmnopqrst
uvwxyzß
äöü &&%*†
(:!?;-—,,"">>''/

355 *Sans serif. Designed by Harald Brödel, 1968.*

ABCDEFGHI
JKLMNOPQ
RSTU
VWXYZÄÖÜ
&123
4567890

356 *Condensed sans serif. Designed by Harald Brödel, 1968.*

abcdefghijkl
mnopqr
stuvwxyzß
äöü
&%*†»«[]§
(:!?,;.−/.."'´)

ABCDEFGHIJK
LMNOPQRSTU
VWXYZ

1234567890

357 *Condensed sans serif. Designed by the author, 1967.*

abcdefghijklmn opqrstuvwxyz Variante

(.,:;!?&%'»«*)

On pages 164 and 165:

358 a and b Sans serif roman and Cyrillic lettered with a round-ended nib and retouched. Designed by the author, 1965.

ABCDEF
GHIJK
LMOPR
STUV
WXYZ

Figure 358a

БГДЖЗ
ЙИЛОП
ФХЦЧ
ШЩЦЬ
ЭЮЯ

Figure 358b

CYRILLIC

Experts differ on the exact history of the Cyrillic alphabet, but this much is certain: the earliest document dates to the ninth century; in the thirteenth century Cyrillic was securely established in the East Slavic sphere of influence. The majority of forms came from the Greek majuscule uncials of the ninth to eleventh centuries, the rest from Greek ligature or the ancient glagolithic alphabet. The oldest form of the Cyrillic is called Ustav (uncial). An early, less geometric development of it is the Poluustav (half uncial). During the sixteenth century a late form of Poluustav was the basis for typefaces. The fifteenth century saw the emergence of a script with budding ascenders and descenders. Cyrillic was generally written with a wide-nibbed pen for a stark thick-thin contrast. In 1708 Tsar Peter I ordered a reform of Cyrillic inspired by the Dutch baroque styles. Both type and hand-lettering were affected. The so-called bourgeois style remained unchanged, except for minor points necessitated through evolving grammar.

Figures 359 and 360 show attempts to write Cyrillic letters in a manner that is, contrary to common practice, historically compatible with the Latin scripts of the Renaissance.

359 Cyrillic capitals. Study by the author.

360 Cyrillic lowercase. Study by the author.

On page 167:

361 Top: Cyrillic sans serif. Designed by Wassil Barakow. From Wassil Jontschew, Die Schrift durch die Jahrhunderte (Lettering through the centuries), Sofia, 1964. Bottom: Alphabet of Cyrillic capitals in the style of a pen-lettered roman. S.B. Telingater, Moscow, 1958.

АБВГДЕЖЗИЙКЛМ
НОПРСТУФХЦЧШ
ЩЪЬЮЯ
абвгдежзийклмноп
рстуфхцчшщъьюя
1234567890

АБВГДЕЖЗИЙКЛМФ
ХЦЧШЭНОПРСТУ
ЩЮЯЫЪ

Figure 361

362 *Calendar page by Walter Breker.*

CHAPTER 4

Type and Lettering in Practice

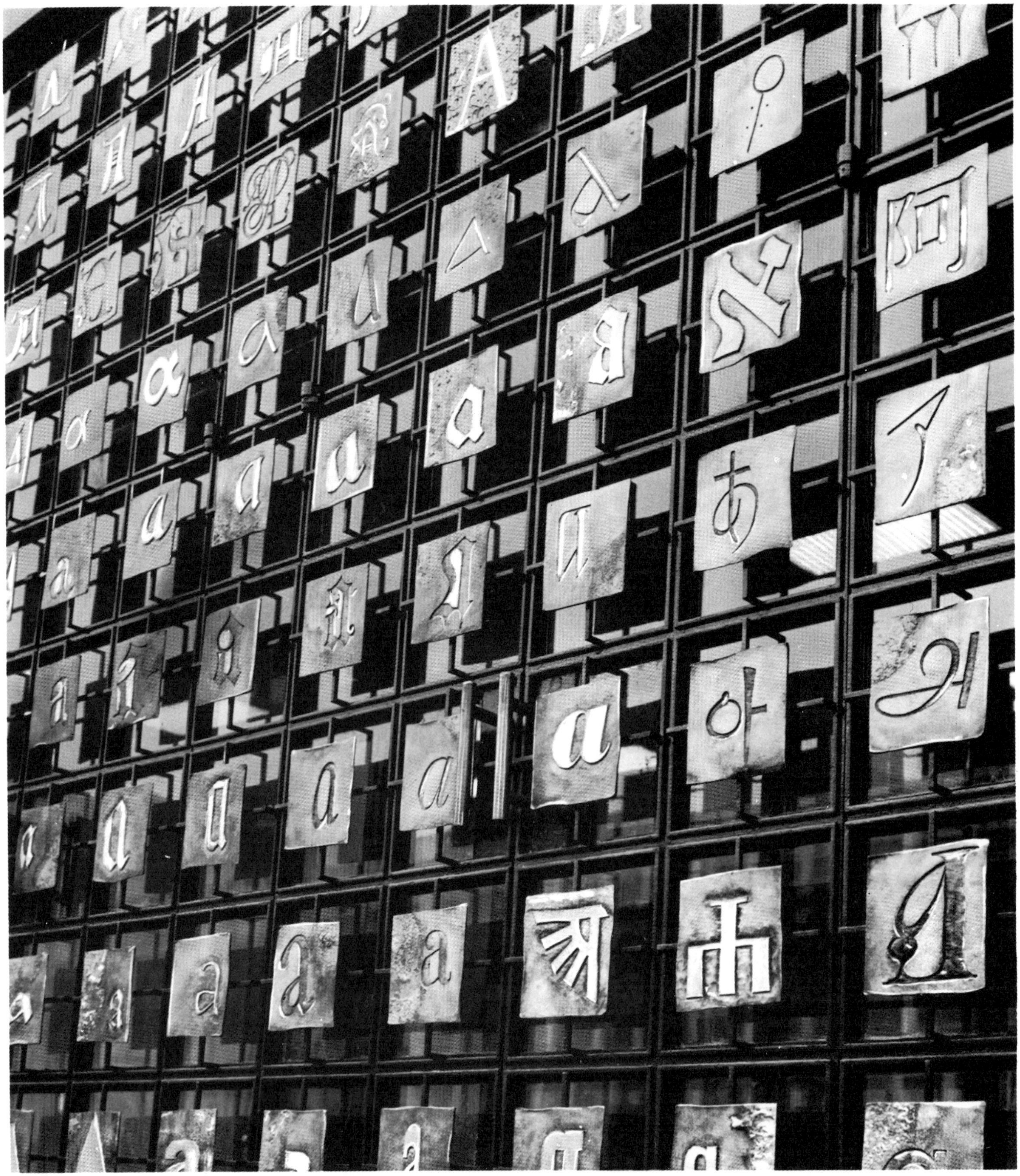

INTRODUCTION

There are a great number of practical applications for type and lettering, including many areas of graphic art, architecture, and crafts. Even the appearance of our streetscapes is affected. The designer has to choose among materials, techniques, typefaces, and layouts for each individual situation. The final decisions depends on his or her skill, experience, and personal taste, and not least on the available materials and circumstances. No rules can be followed mechanically. A proverb says, "From simple matter comes the wisdom of the simpleminded." Our space is limited, and we cannot consider all possible situations.

This book is not an applied graphics or a typography text. Lettering has many more applications than can be discussed here: they range from postage stamps to posters, from birthday cards to utilitarian signs, and let us not forget the calligraphic elements that adorn books on typography. The last section of Chapter 4 is by no means complete. I have discussed only applications where the printed or drawn letters not only visualize a message but are the prime medium of design. We touch on the logotype, the poster, packaging, and the book jacket only in their connection with the written word. This book was not intended as a replacement for the necessary studies that every designer who uses type and lettering has to undertake, nor does it concentrate in any detail on special techniques, such as lettering on plaster, unless the technique is essential to the design process. The information given here is intended only as a jumping-off place: further study and practice is necessary for mastery.

363 *Gate for the door of the Berlin City Library, using the letter A. Designed by Fritz Kühn. (Photo by Fritz Kühn.)*

Typography

Choosing Type[1]

Whether lettering is the sole vehicle of design or just a supplement, whether it is drawn, set, otherwise copied, or created in a three-dimensional form from any appropriate material, the objective is always a useful, harmonious, and lively relationship between text and letters, content and form. No single alphabet can serve all purposes at once. Design elements such as size and proportions of the format, margins, line spacing, the color of paper and ink—all these help determine the end result. Even a relatively neutral typeface such as a sans serif can appear in a rich variety of forms simply through the different arrangement of the letters on the page.

If you want to find just the right typeface, one that fits the purpose and matches the general concept of the work, you, the designer, need the skill to recognize an alphabet's scope for variation and its range of expression. You have to be able to utilize techniques and materials in the best way. Step-by-step instructions are impossible to give. The following recommendations attempt nothing more than to keep the student from making the most common mistakes. A personal and satisfying sense of style will develop over time.

Always consider the individual contributing factors in their relationship to the whole.

Match style and material. Tools and materials contribute significantly to the final form of your work. By the same token, you should never force a style into a form that has developed from an unrelated process. Some tools are only useful for letters of certain sizes. To enlarge pen strokes beyond a certain height or to model three-dimensional forms after fluid brush strokes is to invite disaster.

Consider the attention of the reader. The size, legibility, and expression of type styles have already been discussed. Decide whether these factors should be given equal importance, or whether one of them is preeminent in any given case. The purpose of your work should suggest to what extent you can make the typeface itself the carrier of the intended message.

Where it is necessary, as in advertising, go beyond the simple requirements of visibility and legibility. Try to create a layout that is original enough so that viewers will remember it and recognize it when they encounter it again.

Always keep the entire graphic context in mind. When you choose one typeface, think about the other faces that will be combined with it. Size and weight relationships are subject to the same laws as other compositional elements.

In connection with illustrations or photographs, letters usually occupy a position of minor importance. The inherent contrasts can be heightened by the choice of type. Relatively neutral faces such as Garamond, Walbaum, Bodoni, or the sans serifs are best suited for the job. Sometimes, however, it is possible to create a special unity between image and lettering by using matching materials or tools—for example, brush-drawn letters for a brush drawing. If you are complementing someone else's work, proceed with great caution and sensitivity.

Wherever possible, match the typeface to the content and the spirit of the text. Part of the communicative function of writing is the emotional impression it makes on the viewer and the associations

---

1. Many valuable tips for this chapter came from the book *Schrift muss passen* (Just the right letters) by Leopold Nettelhorst, Essen: Wirtschaft und Werbung, 1959.

364 *Logo for P.N. Stein, a company that makes box springs. Designed by Jurgen Förster.*

that it creates on a conscious or unconscious level. Some designers try to create the desired associations in misguided ways: the name of a coal mine should not be represented by letters that are forced into the shapes of black lumps, nor should a sign for a refrigeration company sport letters that look like melting icicles. Letters do not represent things; where historical examples for such attempts exist, they are nothing more than curiosities. A possible exception is the creation of logotypes, where a fusion of letter and pictogram is justified (Figure 364).

If you choose a typeface for literary texts, for topics of art, music, or history, match the time periods of your media. The typeface will gain expressive force from stylistic coherence. Study and compare the historical styles of various epochs to sharpen your skills. This approach is not always the only possible one to find the right form for a given text. As Paul Renner wrote, "It is not our job to outfit each literary idea with a costume of its period; we only have to ensure it a fitting dress in the style of our own time."[2]

Historical associations are important, but it is also necessary to consider the graphic values inherent in the letters themselves and the resulting psychological effects. Materials and tools possess specific properties, such as smoothness, precision, heaviness, or roughness. The width of the basic stroke and its relation to interior spaces and letter height create narrow or wide, thick or thin letters (Figure 365). These proportions and the thick-thin contrast make the text appear airy and light, thin and dim, or heavy and dark. The direction of the basic strokes and connections between them also create impressions: round, soft, flourished, tight, pointed, hard, stiff. The rhythm of movement can be smooth, flowing, intense, swinging, rigid, controlled, or monotonous; the dynamic can be forceful, energetic, and vigorous or restrained, inhibited, and obstructed. The general expression can be muted, intense, individualistic, differentiated, manneristic, or artificial. It is, of course, impossible to assign a set of descriptive attributes to each alphabet, and it seems obvious that all characterizations are relative. We all connect certain mental images to words, and it is therefore possible to express properties inherent in a thing by means of writing.

The designer's personality, temperament, taste, imagination, and creative power will necessarily influence the outcome, especially if he or she also designs the alphabet. A good graphic artist will never try to be original at any cost. Some jobs require great restraint, and designs for packaging or architectural inscriptions call for almost total suppression of the artist's personal preferences.

Mixing Type

In typography as in calligraphy, it is wise to avoid combinations of more than three different elements in one design. A possible exception is a purely decorative treatment that is often used in contemporary advertisements.

Mixtures are possible in the following situations:

1. Within one alphabet group—that is, mixtures of larger and smaller letters that are otherwise the same.

2. Within a family of styles: mixtures of roman and the related italic, upper- and lowercase letters of an alphabet, or bold and semibold roman.

---

2. Emil Ruder, *Die richtige Schriftwahl* (Choosing the right typefaces), in a memo of Linotype GmbH. Berlin and Frankfurt/Main, Volume 56, November, 1962.

Figure 365

3. Within a historic style: mixtures of
   a. Renaissance roman and fraktur
   b. Schwabacher and textura
   c. Modern variations of Renaissance roman such as Tschörtner and modern modifications of textura
   d. Egyptian or sans serif and English scripts
   e. Historical fraktur and Schwabacher
   f. Neoclassical roman and neoclassical forms of fraktur.

4. In a combination of condensed sans serifs, Egyptian, or modern styles that are written with a brush, and certain modern roman, sans serif, and neoclassical styles.

It is also possible to create a mixture of contrasts, as in the combination of sans serif and Garamond or English script and sans serif.

For decorative capitals consider the style of the alphabet. Ornamental creations that are based on neoclassical form match neoclassical types.

The following combinations are to be avoided:

1. Historical alphabets and their modern variations, such as Akzidenz-Grotesk and Helvetica or Garamond and Tschörtner.

2. Different types of fraktur.

3. Neoclassical and Renaissance alphabets.

Calligraphy

General Remarks

Good lettering can be more than just beautiful or effective. It can be filled with motion, vitality, and the spirit and wit of the topic, but the designer needs to be in command of a wide arsenal of forms and shapes and sensitive in their applications. There is no substitute for

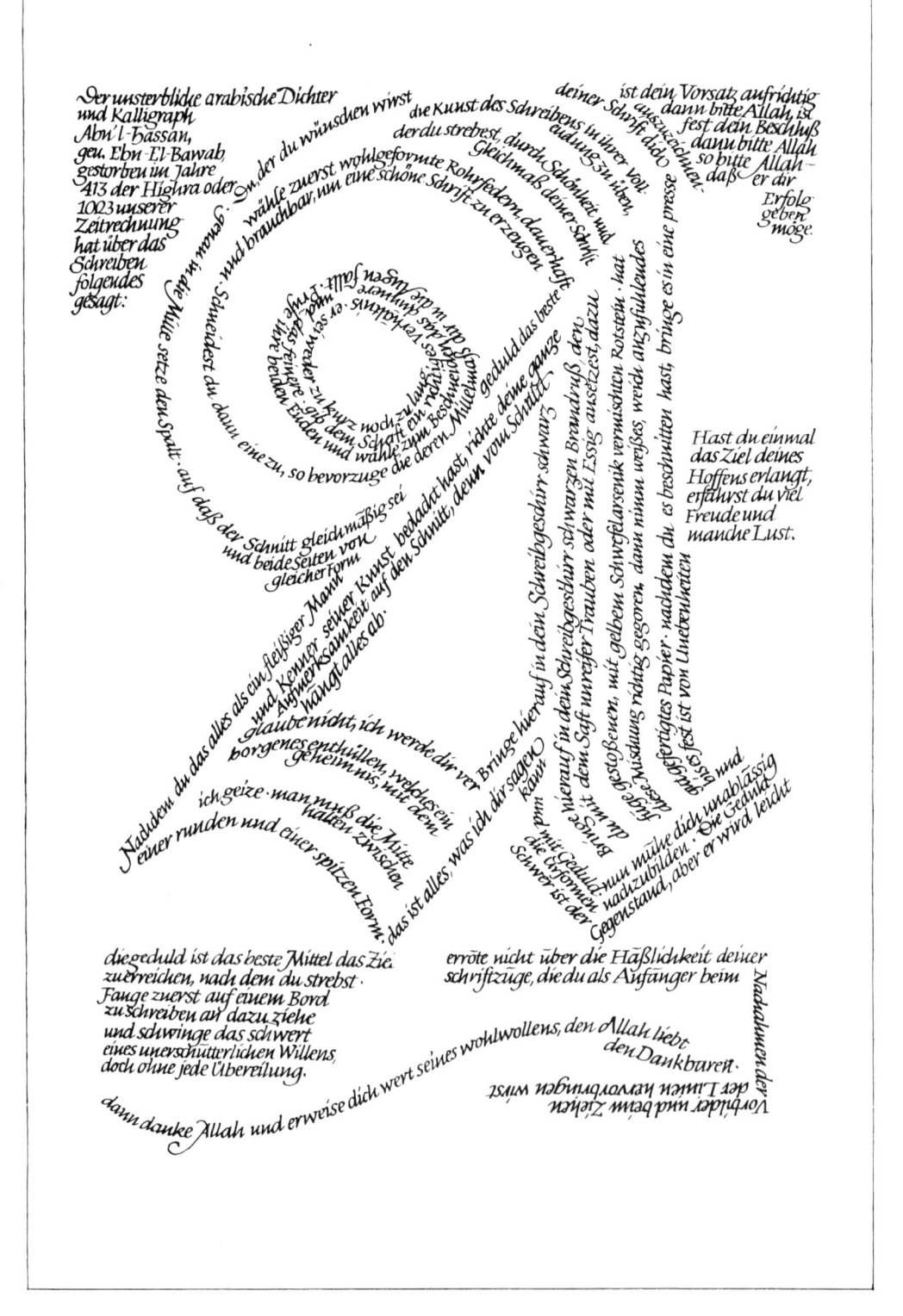

366 *Calligraphic study by Albert Kapr.*

the serious work that a composition requires. The outline, contrasting thick and thin lines, large and small, straight and round, simple and complex structures, horizontals and verticals — all these elements give life to the written word and help create the impression of a balanced whole. The shapes within the x-height carry the ascenders and descenders in an organic process. Quick strokes create interesting accidents and every writing material has inherently different possibilities. Use brushes, steel pens, reeds, chalk, and paper to their best advantage. Unplanned but pleasing effects can be combined by cutting and gluing, but the result should always look as if it was produced by one stroke.

Documents and Short Texts[3]

General Remarks
Documents such as diplomas and certificates and short texts such as speeches are frequent subjects for calligraphic work, and many techniques are available. You can write on just one side of a single sheet and frame it or store it in its own portfolio. You can present it as a scroll, or fold and box it. Works of larger scope are best bound in book form. In each case text and surround should be in unison, and binding techniques should be taken into account from the earliest planning stages on (see pages 187–190). The grain direction of the paper is important if the paper is to be folded or scored later on. A professional bookbinder can be of help if the project requires specific materials, but get advice before you start your work.

It seems redundant to point out how important it is to plan carefully. Choose several appropriate letter styles and write the entire text in each one. Letter special sections such as headings, initials, or other important parts in more than one size and color to create a pool of images from which you can later choose. Most purposes are served well with a more or less representative selection of roman capitals and lowercase letters and cautious use of italics and decorative elements. Forms with strong historical associations such as fraktur or uncials have a tendency to seem antiquated, even when used as accents, but this effect could be balanced by a decisively modern layout. Use color only to accent the most important word or line of your text, or for initials. The most beautiful effects are often created with black and gold, but red and turquoise can produce excellent results. Mix your red from vermilion and carmine red, add some white, and vary the proportions to achieve a cool hue for use on a bluish paper and a warm one for yellowish paper and parchment. Coats of arms and other insignia look best when they are embossed and gilded.

When all the words have been written, cut them out and rearrange them on your format. Whether you choose a symmetrical or nonsymmetrical layout depends on the requirements of the text. Longer sequences are best served by flush-left text. When you like the arrangement, glue the pieces in place.

Parchment is the material of choice for very important occasions or when durability is a concern. If the additional work is justified, consider the use of gold for especially valuable pieces. Do not take any short cuts, however — imitation parchment and fake metals in paste and powder are bad taste, even if they are used to make nothing more than a diploma for the local ballet-dancing club.

---

3. Much of the information in this section comes from Edward Johnston's book *Writing and Illuminating and Lettering*. New York: Taplinger, 1977.

Besides, the metallic colors will turn black in a short time. Handling the real thing requires great skill and much practice.

Using Parchment

Pure white vellum is preferable to the slightly yellow version, sometimes called "antique," which is often uneven in tone because it is gathered from animals that died naturally rather than from slaughtered ones. Parchment has two distinctly different sides: the inner, lighter one, called the flesh side, and the outer side, usually of a darker, yellowish color, called the hair side. The hair side is often chosen because of its interesting patterns, but if a text requires several pages, you will have to write on both sides. Arrange the pages so that hair sides face each other and flesh sides face each other.

Choose skins with few or no callus spots. A thicker skin can be used for a single document, especially if it is to be framed, but a folded page requires thin and flexible material or it will not fold or remain open easily (see Figures 367 and 368).

Dealers who specialize in calligraphy materials, such as the Scribes Art Shop (568 Jefferson Plaza, Port Jefferson, NY 11776), can supply vellum prepared for writing.

Parchment for bookbinding purposes is glossed with an application of egg white and then buffed to a matte sheen after drying. This prepared surface makes writing difficult, because the ink does not adhere well and often collects in small drops. Similar problems arise in gilding. If you cannot obtain special writing parchment, there are several methods for modifying the surface of treated skins. Try the procedures on scraps first to test the kind and the intensity of the layer in question.

Edward Johnston recommends the use of finely ground pumice stone. Rub it into the surface with a piece of supple leather until you see a slight fuzz, then remove the particles with a clean silk cloth or with a brush, and shake off the rest. If the parchment surface has a desirable pattern, which could be destroyed by scraping, an alternate method is to wash it with a solution of alum. Because parchment has hygroscopic properties—that is, it absorbs water—it has to be stretched onto a board and secured with tacks or glue. This procedure makes it necessary to start with a larger piece than you need for the final work, because the edges have to be trimmed off.

Make a lining tool by assembling a blunt needle and a handle and draw lines gently. Score, but do not crease, the parchment. Setting your light at an angle will make the lines visible. Write on loose, not stretched, parchment and use black Chinese ink in stick form with a little watercolor added. Choose only the best pigments and control uniformity and lightness by adding small amounts of white to your colors. Any area that is to be gilded should be kept clean during the writing process.

Raised Gilding

The technique of raised gilding requires swift action. Keep all tools and materials at hand. You will need gold leaf and a supply of gesso, the ground to which the gold sticks, as well as the following equipment: a hard, totally smooth and clean work surface, pumice stone and a leather cloth to roughen the surface of previously prepared parchment; a hard pencil to draw outlines; pen and brush to transfer the gesso; a breathing tube or 4-inch (10-centimeter) long tube of paper with a diameter of ⅛ inch (7.5 to 10 millimeters); a gilder's cushion; a gilder's knife; a gilder's tip; a sharp pair of scissors; a needle to pierce air bubbles in the

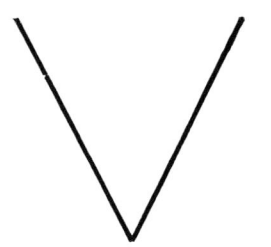

367 *Wrong: parchment too stiff.*

368 *Right: parchment lies open easily.*

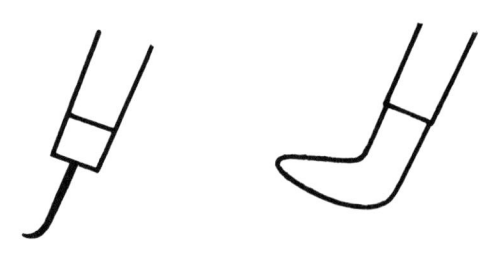

369 *Scriber. (Drawing from Johnston.)*
370 *Burnisher.*

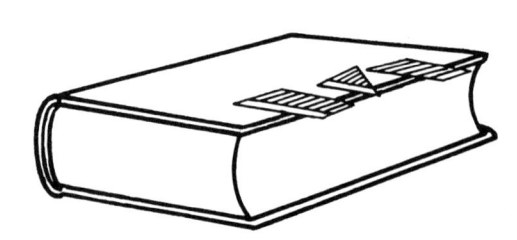

371 *(Drawing from Johnston.)*

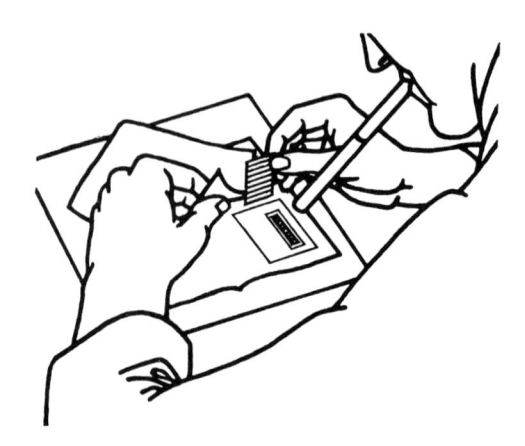

372 *(Drawing from Johnston.)*

gesso; glassine through which to smooth the gesso and to burnish the gold; a small dusting brush to remove the pumice dust and excess gold; a burnisher (Figure 370) or a rounded and smooth bone folder; and a small dish to mix the gesso.

Gesso-based recipes for raised gilding appear in sources as early as the ninth and tenth centuries. The following recipe is adapted from Helmut Hirmer.[4]

- 8 parts slaked plaster of Paris
- 3 parts white lead (NOTE: white lead is toxic. Wear rubber gloves when you handle it, as it may be absorbed through the skin, and avoid inhaling it.)
- 1 part fish glue
- 1 part cane sugar
- 5 parts distilled water
- a dash of pigment (Indian red or cadmium red)

Grind the gypsum, white lead, sugar, and pigment individually with a mortar and pestle. This is a very time-consuming process, requiring 45 minutes per ingredient to achieve an absolutely smooth consistency. Measure out the amount of each powder called for in the recipe, combine them in the mortar and mix them together. Add fish glue and water a little at a time. Do not stir, but carefully work it together. Should air bubbles form, you must prick them open with a needle. It will take another 45 minutes to get the materials thoroughly mixed. If the paste is too stiff, add a few drops of distilled water.

Prepare a piece of cardboard wrapped in aluminum foil and put little "buttons" of the gesso ½ inch to 1 inch (1.5 to 2.5 centimeters) onto the foil to dry. Store in a dry and dustfree place and it will keep indefinitely. To reactivate the gesso, crumble a button into a small glass with a round bottom and add a drop of glair. Let it soak in for 5 minutes. Wearing a rubber glove, stir it with your finger until the paste has the consistency of putty. Cover it with glair and let it stand for 10 minutes more. Mix again until it is the consistency of cream. Open air bubbles with a needle or pin.

To make glair, beat the white of an egg to a stiff froth. Allow it to settle at room temperature for twenty-four hours. You will find a small amount of liquid at the bottom of the bowl: discard the froth and add the liquid to an equal amount of water.[5]

Other recipes may be found in *The Calligrapher's Handbook*, edited by Heather Child (New York: Taplinger, 1986).

Ready-made gessos are no longer available, but acrylic medium—gloss, matte, or gel—can be used successfully. Keep the gesso in a glass container with a lid and use only as much as needed. Well-covered leftovers can be thinned with water until the consistency resembles that of cream. Stir frequently during use and poke any air bubbles open with a needle.

If you used Johnston's scraping procedure for the preparation of parchment, rub the surface to be gilded with pumice powder before you draw your outlines. This will ensure a good contact of gesso and parchment. A particularly greasy spot can be tackled with the blade of a small knife. Then draw with a hard pencil or copy the desired shape onto the designated area. If you washed the parchment with alum solution, treat any difficult spots with a solution of 8 to 10 drops of hydrochloric acid to 3 ounces (.10 liter) of water. Apply the liquid with a brush.

---

4. Helmut Hirmer, "Die Gessovergoldung" (Gesso gilding), in *Das deutsche Malerblatt*, 4/90.

5. Recipe provided by Jerry Tresser, The Scribes Art Shop.

If the paper is not absolutely horizontal on the work surface, the gesso will collect in the lower sections of the letters and the relief will be uneven in height. Outline the form with pen and liquid gesso, fill in with a brush and thicker gesso, and build the relief up within the contours of the letter to a height of at least 1/32 inch (.25 to 1 millimeter) for letters of wide stroke widths. The gesso will shrink as it dries, and it is possible to add further layers, after the surface has been slightly scored with a knife.

Work quickly. Delays may cause an uneven surface, but small problems can be corrected. Never burnish the surface before the relief height is satisfactory: it is difficult to add further layers to a smooth one.

Gesso that has been accidentally placed outside the letter can easily be cut away after it dries; the same is true for small drops of gesso spattered elsewhere on the parchment.

Should the gesso become brittle during work, scratch it off and add glue or binder to it, but frequent stirring will prevent most problems. To predict the length of the drying time is difficult: it depends on the ingredients of the gesso, the thickness of the relief, the humidity, and the room temperature. Make some samples of the same kind as your project and test them periodically for dryness. Thin lines have to be gilded before thick areas. Johnston suggests drying times from two to twenty-four hours.

Gold leaf has a tendency to stick even to surfaces that are not covered with gesso, so it is useful to cut a template from lightweight paper to protect the area surrounding the gesso buildup. Put loose gold leaf on the gilder's cushion and cut it with a gilder's knife; cut tissue-backed (patent) gold leaf with scissors. Do not sandwich gold leaf between two sheets of paper. You can cut several pieces at a time and position them at the edge of a book or a box, as shown in Figure 371, to use as needed.

To activate the stickiness of gesso you have to breathe on it, section by section, through a breathing tube (Figure 372). Move the tube close to the surface, and take care that your breath does not condense into drops that might fall onto the gesso. Protect already gilded parts, as well as those that have not yet been treated, with strips of paper. The warmth will dull the gold, while gesso that gets warmed up too early loses its ability to hold the gold leaf.

Apply the gold immediately after the gesso is warmed. If you use a gilder's knife to cut the gold leaf, you will also need a gilder's tip — a brush used to pick up the cut sections. Touch the tip to the gold and it will adhere easily. Now hold the tip over the spot that is to be gilded and, using a cotton ball, press the gold onto the paste from the back side of the frame. This procedure is not only convenient but also safe.

If you cut the gold leaf on tissue backing, set the gold onto the warmed gesso and, with your fingers, push it down gently paper side up (Figure 372). Then remove the tissue backing. Cover the gold with glassine. Hold it in place with your left hand so that it will not shift during the following procedure and push the gold onto the gesso through the glassine with the pointed end of the burnisher. Then rub the entire surface with the flat side of the burnisher. If you are not sure whether the gesso is hard enough for this procedure, use a tightly rolled wad of cotton instead of the burnisher.

If the gesso dries out before the gold is applied, it will not adhere well. The same thing will happen if you touch the prepared gesso and leave a trace of oil. To remedy this problem, roughen the

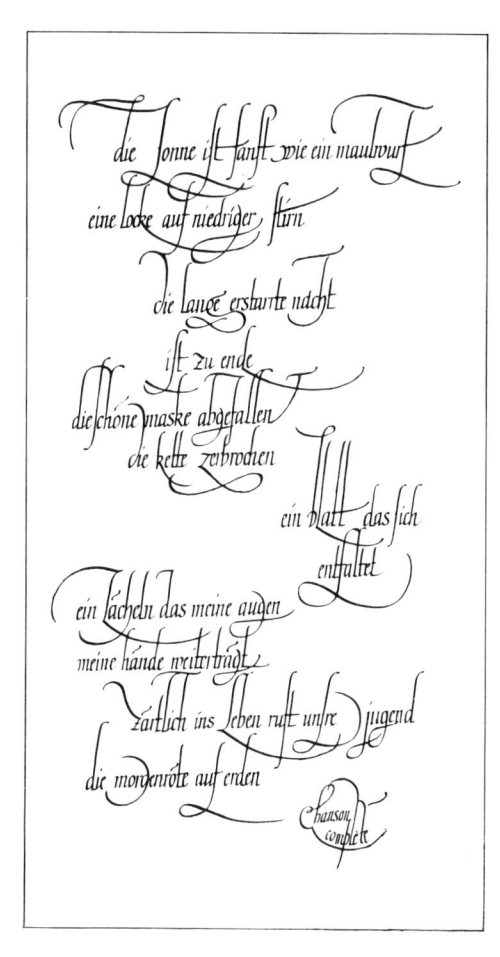

373 *Page (reduced) from a handlettered book of Paul Eluard's poetry,* I am not alone, *by Irmgard Horlbeck-Kappler. Leipzig, 1963.*

surface and apply a fresh layer of gesso or a thin layer of glair, which can also be used as adhesive for small repairs in the gold leaf.

Keep adding layers of gold leaf until they no longer adhere. You may need as many as three or four layers, because gesso particles can penetrate the first layer, especially if it is too liquid or if the gold leaf is very thin.

Do not burnish until you are certain that the gesso has dried thoroughly, and test a sample piece. Homemade mixtures are ready for burnishing just a few hours after the gold application.

Start the burnishing process with round movements of the burnisher, and continue with straight strokes. If you feel any resistance at all, stop immediately, look for any gesso on the burnisher, and remove it. This problem can be caused by too much glue in the mixture or by insufficient drying time. Sometimes it is enough to add more gold leaf and wait a little while before burnishing, but if the second attempt is not successful either, you have no choice but to scratch off the gesso and start anew.

Small imperfections in the layers or in the burnishing become clearly visible if you use a strip of white paper as a reflector. Protect your work from the humidity in your breath, and never touch it with your fingers.

Excess gold leaf is usually easy to remove after burnishing is completed. If you cannot clean the pieces away with a brush, scrape them off gently with a knife. Burnish again after eight to ten days, and once more two weeks later.

The Handlettered Book

General Remarks

Since the advent of printing, its ever-increasing popularity has displaced the handlettered book as a means of mass communication, education, and entertainment. Today, the raison d'etre of the handlettered book lies in the realm of bibliophilia. Each piece is produced individually, though it is possible to reproduce a given work in small editions by way of lithography or offset printing.

For more or less official occasions, speeches, company histories, and the like can be presented as handlettered volumes and make appropriate gifts to honored guests as souvenirs of the occasion. A well-executed calligraphic design can meet such requirements more efficiently than a typeset book, since the small number of copies needed would rarely justify the necessary time and expense for commercial typesetting and printing.

The importance of the handlettered book for private use should also not be underestimated. There can hardly be a more personal or unique gift. For many of these works it would be difficult to find a publisher, but it is possible to create a small library of handwritten books for specific purposes. For example, Goethe's father collected his son's early poems in this form.

A third group of handlettered books consists of artists' books—works created by painters or graphic artists who are less interested in the official or private motive for the text, but rather wish to interpret the contents in their own way. This can be done using lettering alone, but more often the text is combined with illustrations, and the best results come from works in which the character of all elements is carefully matched—for example, the works of Matisse and Picasso shown in Figures 473 and 474, page 221.

The earliest printed books were imitations of handwritten ones. Today the handlettered book should not seek to emulate the printed book, but it is useful to take into account the rich experiences of typesetters, not just where the strictly technical aspects are concerned but especially the considerations of logic and aesthetics. For a handwritten book issues such as format, decoration, letter size, and arrangement of text blocks are less constrained by the demands of production techniques than for a typeset book. The designer has much more influence on the final product and can express his or her personality in many ways, freely and unencumbered by many restrictions. Different occasions — festive speeches, works of poetry or prose, and so on — demand specific forms appropriate to their content. Legibility is an issue of varying importance, as expressive poetry may be better served by a lettering style of comparable emotional value at the expense of clear presentation of each letter. Such variations and decisions are impossible to make in a typeset volume. Calligraphy can be used to interpret and visualize text in a way similar to illustration.

A series of planning steps is necessary for the production of a handlettered book, and these will be discussed in the following sections.

Format and Page Layout

If you are planning to use paper for your book, you need to determine the grain direction and the size of your sheet before you decide on a format (see the section on Paper Structure, page 187). There is an important difference between layout for a single loose page and for a book: the single sheet stands alone, but a book opens to a spread — a pair of facing pages that have to function as a unit.

Every page layout has to be related to the layout of the facing page. If the text block were placed exactly in the middle of a page, surrounded by margins of equal width, the spread would optically fall apart in the middle. Jan Tschichold wrote, "Harmony between page format and page layout comes from equal proportions between both of them. If a union is achieved between layout and format, margin proportions become a function of page format and the construction process. The elements are interdependent. The margins do not dominate the page, they develop naturally from the page format according to a form canon."[6] Tschichold has reconstructed such a canon, which is the basis of many documents and incunabula of the late Middle Ages (Figure 375).

Tschichold found other ways to create the same proportions. One is to divide width and length of the page into nine equal parts (Figure 376); another to arrive at the nine parts with the help of Villard's Figure (Figure 377).

The nine-part division gives good results even with other page formats. Examples are 1:1.732 (the golden section — 21:34), 1:1.414 (the international A format), 3:4, 1:1, and 4:3 (Figure 379).

Tschichold considers the nine-part system the most beautiful, but not the only possible one. Referring to some of the illustrations in his essay, which are reproduced here, he further wrote, "A twelve-part system creates a larger text block, as we can see in Figure 16, ... Figure 17 shows pages with the side proportions of 2:3 divided into six equal parts. Along its length the paper may be separated into any number of parts if necessary. Margins even narrower than those in Figure 16 are possible; the only

---
6. Jan Tschichold, "Papier und Druck" (Paper and printing), in *Typografie*, Vol. 3, 1966.

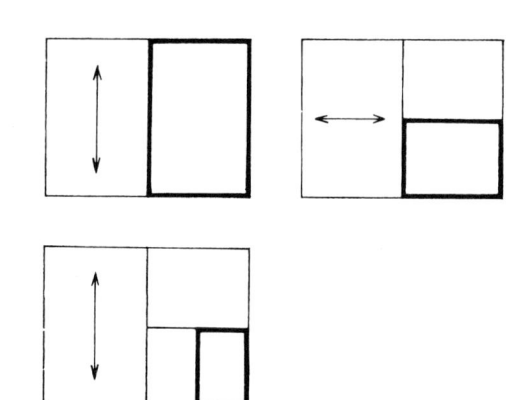

374 *Page formats, showing grain direction of the paper: folio, quarto, octavo.*

375 *Jan Tschichold identified the plan of many medieval handwritten texts and incunabula: the proportions of the page as well as of the text block are 2:3. The height of the text block equals the width of the page. (Figure 5 of Tschichold's essay, "Willkürfreie Massverhältnisse der Buchseite und des Satzspiegels".)*

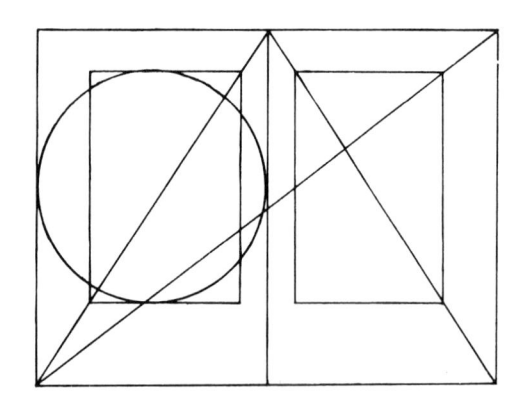

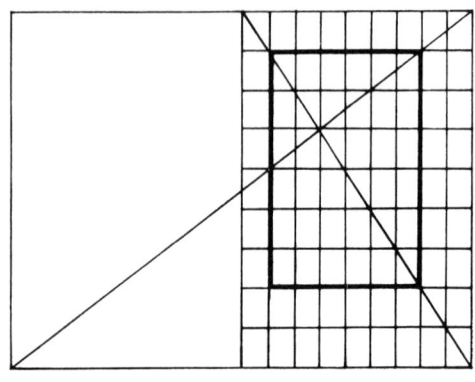

376 *Rosarvio's construction. Like Tschichold's, it assumes page proportions of 2:3 but uses a system of nine basic units. The result is the same. (Tschichold, Figure 6.)*

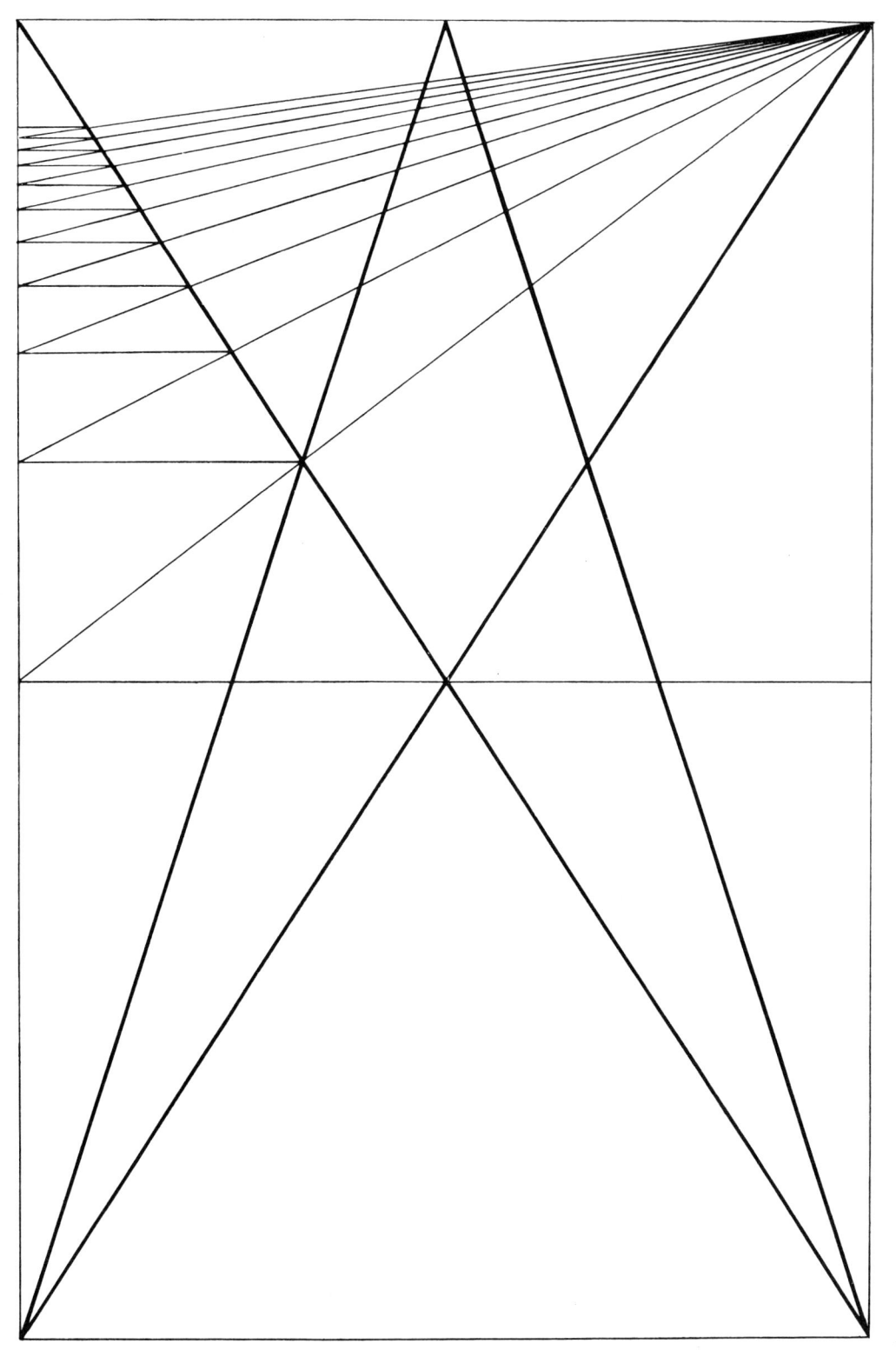

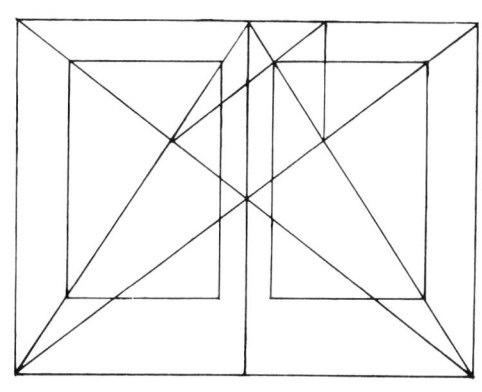

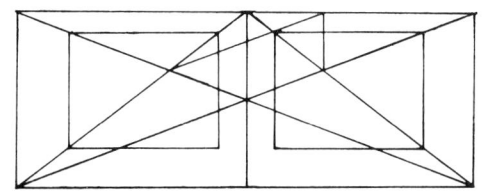

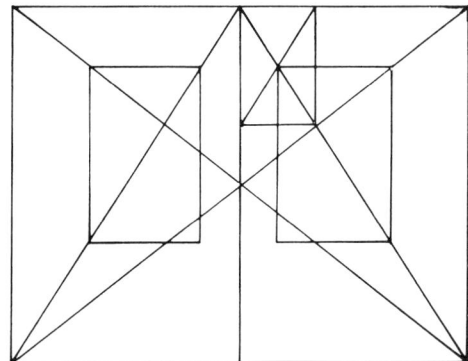

377 Villard's figure. The heavy lines make it possible to divide any other line into any number of segments. No rulers are necessary. (Tschichold, Figure 9.)

378 It is possible to use Villard's system to arrive at the nine-unit plan; page proportions remain 2:3. (Tschichold, Figure 8.)

379 Height and width are divided into nine parts. The proportions are 4:3. (Tschichold, Figure 15.)

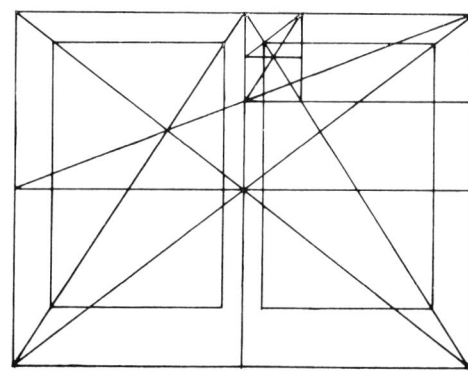

380 Villard's system used to divide the page into twelve parts. The proportions are still 2:3. (Tschichold, Figure 16.)

381 Division of the page into six parts, with the same proportions. (Tschichold, Figure 17.)

181

requirement is that the connection of text block and single as well as double page has to remain intact through the diagonals. The diagonals guarantee a harmonious placement of the text block on the page."

Other examples of good placements and proportions can be found in Albert Kapr's layouts, as shown in Figure 382. Kapr is somewhat more tolerant when it comes to the requirement of aligning the lower right and the upper left corner of the text block with the diagonal of the page. Especially when he designs for narrow formats, he finds deviations from this rule advantageous.

The classic proportions are not the only possible ones. The handlettered book can benefit from extremely wide blocks (Figure 383) or extremely narrow ones. Even asymmetrical arrangements are right for certain situations.

Beginners usually depend on models and examples, but the experienced designer will find his or her trained eye a more reliable guide. In determining the size of the margins it is important to consider that approximately 1/16 inch (3 millimeters) will be trimmed off the top, bottom, and outer sides of the page during the binding process. Binding will also reduce the width of the inner margins by about 1/32 inch (1 millimeter), depending on the type of binding process used.

Close relationships exist between margins, line spacing, line length, and letter size. In a justified text, the length of the line is identical to the width of the text block, but in a large format the text can be arranged in two columns (Figure 384).

Limit the number of words per line in a handlettered book to eight to ten. This determines more or less the height of the letters: you will establish the precise dimensions through several trials. If no specific page size is given, you can also

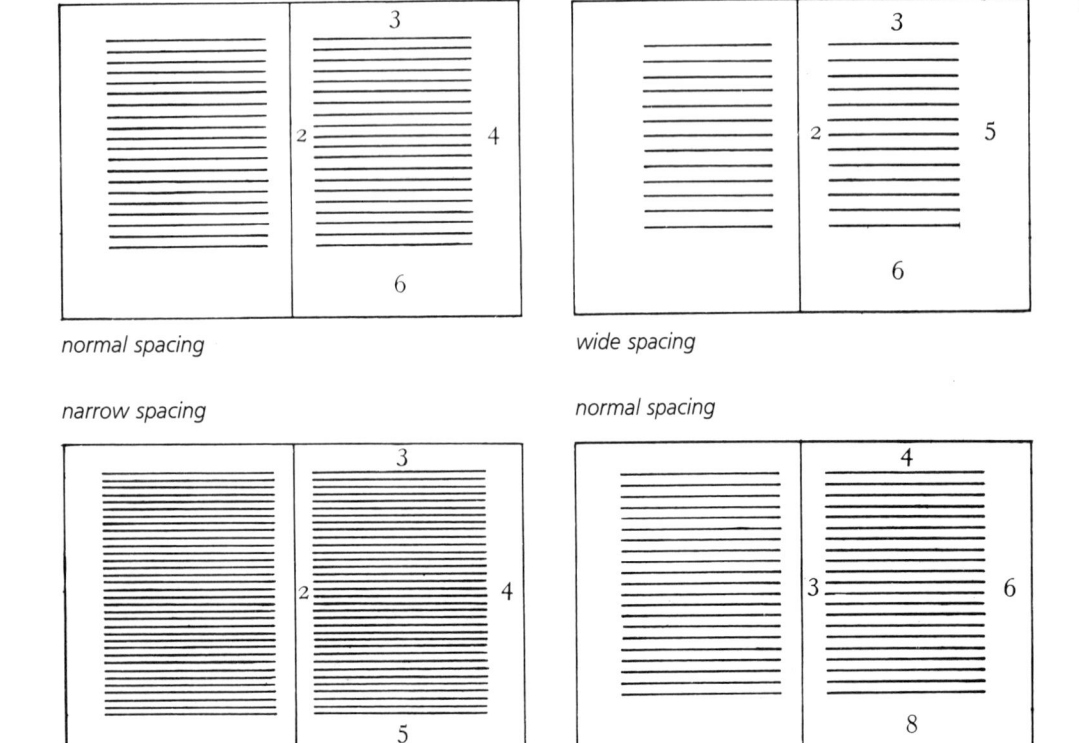

382 *Page layout variations by Albert Kapr.*

normal spacing

wide spacing

narrow spacing

normal spacing

reverse the process, start with a particular size letter and deduct the line length, line spacing, and all margins from this unit. It is often necessary to combine both approaches in one project.

Another factor that has impact on the margin width is line density. Loosely spaced lines require wide margins, while dense layouts need less space around them. Allow large letters narrower margins than small letters.

Designing the Text Block[7]

Chapter Openings and Endings. Many different design elements can be used, but within one book there should be continuity. The most common way to differentiate a chapter from the rest of the text is to start it lower on the page. (The distance from the top of the page to the first line is called sinkage.) The white space can occupy as much as one-fifth to two-thirds of the entire text block.

Headings and initials are other important design features of the chapter opening. Align the headings with the text or the top of the text block, and leave a little extra space to the beginning of the text. The headings can be centered or flush left; the latter is the most frequent choice for handwritten texts. If the text has subheadings, leave three to four line spaces between blocks and place the subheading somewhat below the middle in the empty space. According to Kapr, "The headings represent the fusion lines of the text. They are the skeleton to which tendons and muscles of the body are attached." Size, color, and letter style for headings are discussed in Chapter 1 and at the beginning of this chapter.

Past centuries brought us a fascinating and ever-changing variety of initial capitals. Today we still use ornamental initials, but with restraint when it comes to size and decoration. Choose a style that harmonizes with the rest of the text, and refer to Chapter 1 for more details.

If both the first and the last column of a text appear on the same spread, the final column should preferably be longer than any white space at the end of the first page. Arrange the columns of the first page accordingly: you may have to make several layouts to solve the problem.

Paragraphs. Indents mark the beginning of a new paragraph and make it easier to read the text. In a normal-sized line indent two or three letters, in longer lines indent more. Indents are of special importance in handwritten books since the text is usually not justified, which makes paragraphs hard to discern. Line spaces can separate paragraphs, but tend to disintegrate the text block and may look worse than a somewhat irregular left edge.

The first paragraph of a chapter or a paragraph after a subheading needs no indent because the preceding line space separates it from the rest of the text. A flush-left beginning of a new paragraph is also easier to reconcile with a heading. It is, of course, possible to ignore this rule for special design purposes. You could start paragraphs with colored initials; if they are small they can be part of the line, but they can also occupy space in the margin area. Do not overlook paragraph marks, even though they may seem a little old fashioned (Figure 385).

Page Numbers. Since the handwritten book is likely to be used in ways different from conventional books, page numbers, or folios, are not strictly necessary, unless the volume is very thick. Page numbers are most frequently placed

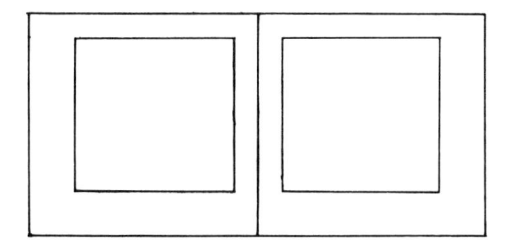

383 *Very wide margins.*

384 *Page layout with two columns of text.*

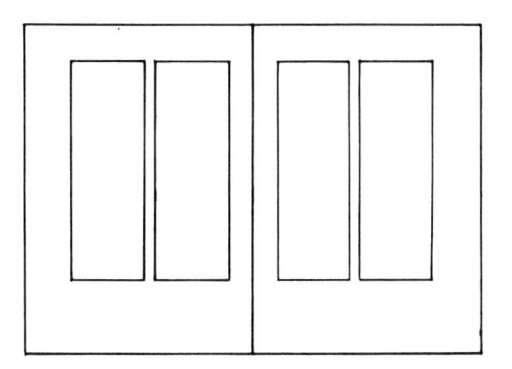

---

7. This section is based on material from Albert Kapr's *Buchgestaltung* (Book design). Dresden: VEB Verlag der Kunst, 1963.

below the text block at the outer margins or centered. Both symmetrical and asymmetrical layouts can be numbered this way, but the folios can also be positioned at the left margin on both sides of the double page. Do not place both numbers in the inner margin unless the entire spread is symmetrical. The numbers can be emphasized by their position or by color, but it is best not to give them too much prominence.

Marginalia. Old documents frequently contain remarks that were added later in a different hand in the margins. These notes often grace the page and suggest the possibility of including them intentionally in the design. Marginalia, also called side bars, should be lettered in a smaller hand and in a less formal style than the main text. Use them for less important features such as references to other parts of the text, subheadings, page numbers, or footnotes. Roman text could be supplemented by italic marginalia, black ink by marginalia in a contrasting color such as red or blue. Start the first line of the marginalia in the same line as the text part that it refers to. The following lines cannot be matched, since the smaller size of the lettering requires smaller line spaces (Figure 387). Consider the width of the margin and fit the side bar to it: about half the width of the margin is a good width for the side bar. Letter all marginalia on separate sheets of paper to determine the right proportions before you add them to the text.

Footnotes. Footnotes are text supplements. They tell the reader about source material or other related facts, and are placed at the foot of the page or at the end of the chapter. (In a handlettered book, the same information can be given in the form of marginalia, as already mentioned.) The need for footnotes will not arise often in calligraphic books, but some hints for their arrangement may be helpful nevertheless. Mark the appropriate spot on the text with a small raised numeral (the size that you would use to write fractions) in a style that matches your text alphabet. Put the numeral right next to the word that needs further explanation, or after the punctuation mark at the end of the sentence, if the footnote refers to the entire statement. If a page contains no more than three footnotes, you can use asterisks and other traditional symbols such as the dagger instead of numerals (Figure 386). The space that the footnotes occupy has to be integrated into the text block. Shorten the column by the appropriate number of lines and separate the text from the footnotes by one line space. If the resulting space seems too wide, fill it optically with a line, drawn the full width of the text, using the same pen that was used for the text. Letter the footnotes themselves considerably smaller than the text and start each one with the appropriate numeral or symbol. The number should be the same size as the footnote lettering, and written on the baseline, followed by a period or a space. You may start the footnote text flush left or indented. If the indents of the text itself were three letters or ½ inch (1 centimeter) deep, use the same measure for the footnote indents. Start a new line for each footnote.

Poetry. "The perfection in the language of poetry has to be highlighted optically. The typographic expression follows the strict laws of the poem's meter or its free rhythm," says Albert Kapr.[8] This concept, though meant in relation to typog-

---

8. Albert Kapr, *Buchgestaltung* (Book design). Dresden: VEB Verlag der Kunst, 1963.

385 *Paragraph marks.* **386** *Asterisk.*

Figure 387

hamonohamonohamon
hamonohamono
monohamonohamon
ohamonohamonoh
hamonohamonohamo

onohamonohamo
ohamonohamonoha
monohamono

184

raphy, is even more relevant for the handlettered book. Here, the graphic artist has considerable more leeway than the designer of a typeset book, but there must never be a discrepancy between the form and the original intention of the author. Lines and stanzas have to be carefully observed, sonnets and terza rima must remain just that.

Since the lines of a poem can be of quite different lengths, pick a medium one as measure for the width of the text block, and let longer lines protrude into the margin space. Separate extremely long lines into two parts and start the second one after a deep indent in a new line. If this should be necessary more than once in a poem, all indents should be the same size. Sometimes the ensuing layout is less than pleasing, especially if there are lines that contain only one word as a result of the separation. In this case it is better to write the word at the end of the following line and separate it with a vertical line, rather than isolate it. Two or three words in a group are preferable to single words, and repetitions of these right-end additions to lines should also be aligned with one another.

Titles are best arranged flush left with the text. Allow one line space between stanzas and at least that much between the title and the first line of the poem.

Front Matter and Colophons

Dating back to the invention of the printing press, the front matter contains all the publishing information about a book. It comprises all the material before the text begins. In older, handwritten books the text started on a recto (right-hand) page with an ornate initial, or with the first words or lines highlighted by a larger size or different color. Some texts used both initial caps and other ways of stressing the beginning. The title of the book could be written in small letters at the beginning of the text, while all other information appeared in the colophon at the end of the book.

Title pages of the sort that we know from printed books are also found in calligraphic ones because they are aesthetically pleasing as well as informative and provide a kind of overture for the following text. The handlettered book, however, offers a wider range of variations than the printed book for the design of the title page.

The traditional elements and sequence of the front matter is as follows. Page one, the first right-hand (recto) page, is usually a half-title page, containing only the title of the work. It may be replaced by a dedication or left out altogether in the calligraphic book. Page two, the first left-hand (verso) page, is either blank, or contains a frontispiece or a series title (if the book is one in a series); in the calligraphic book it may be blank, or it can be incorporated into the design of page three, the title page, which contains the title and the name of the author. In the printed book the title page carries the name of the publisher, and sometimes the city and year of publication. In our handlettered book this page may give the name of the letterer as well as the place and date executed. If the book contains material taken from another work, include the name of the original work and its author (and translator, if applicable), on the title page. Permission to use copyrighted material must be obtained from the copyright owner, and appropriate credit for it should be given. Page four contains the copyright notice, if copyright protection is desired. A dedication may be placed on page five. Page six is blank, if a dedication precedes it. Handlettered books rarely need a table of contents, but the contents may be placed on page seven or, if there is no dedication, on page five.

The text proper, if there is no preface, begins on page nine, with the left-hand page (page eight) blank.

To sum up the arrangement of front matter for a calligraphic book:

page 1	(recto)	blank, half-title, or dedication
page 2	(verso)	blank
page 3	(recto)	title
page 4	(verso)	blank or copyright notice and credits
page 5	(recto)	dedication or table of contents or beginning of text
page 6	(verso)	blank, if a dedication or table of contents precedes it
page 7	(recto)	beginning of text

Traditionally, printed books end with a colophon, which contains facts of production: the book designer, the typesetter, typeface, printer, and other details of manufacture. If the book is a limited edition, its total run and the number of the individual copy may be placed here or in the front matter.

The title page is an important element of the book and its design should complement that of the text. For example, align the first line of the title page with the first line of the text block, and the last line with the bottom line of the text block. The basic lettering style of the title should relate to the text. The title itself, that is the name of the book, as distinct from author or artist, can be emphasized by increased size or stroke width or by choosing a contrasting letterform. If you choose a contrasting form, echo it in the text design, for example, as initial caps or for headings.

Two or three variations in the front-matter lettering should suffice. It is not necessary to graduate the letter size according to the importance of the text, and the colophon can be either the same size or slightly smaller than the main text. It is more important to balance height and width proportions of the elements on the title page, stressing the upper section over the lower (see Figures 390 and 392). The letters can fill the entire page in a decorative layout, but they should not overwhelm the text in the main part of the book.

Consider the relationship between lettering and the surrounding space of the title page when you combine long and short lines. Let the shortest line follow the longest one and vice versa to avoid a pyramid or stair-step layout. Figure 388 shows four classical arrangements of three-line blocks.

Whether you choose a symmetrical (centered) or asymmetrical layout for the front matter should depend on the text pages: all primary elements should follow the same design principle, while secondary elements may deviate from the rule. Figures 389 and 390 show examples. Figure 391 shows a poem laid out flush-left, but a different layout would have been possible; the title and text block could have been centered, headings and page numbers asymmetrical. Figure 392 shows an asymmetrical layout.

Consider text blocks, titles, and illustrations primary design elements, headings and page numbers secondary ones.

Planning the Layout

To get an idea of what the page will eventually look like, cut blocks representing the text on a spread from gray paper and arrange them on the pages. This will allow you to make adjustments before you letter the final piece. Even after a satisfactory layout has been achieved, it may be wise to letter several practice pages with slight modifications. This effort, time consuming as it may be, is usually justified by the importance

388 *Classical three-line layouts. (After Kapr.)*

389 *Asymmetrical layout of heads and folios with symmetrical text.*

390 *Centered title layout.*

391 *Asymmetrical arrangement of text, heads, and folios.*

392 *Asymmetrical arrangement of heads.*

of the occasion for which you are creating the calligraphic book. Next, rough out a complete version of the book, including all headings, initials, and other special features. Cut the lines out and arrange them according to your gray paper layout. Glue the strips down using double-sided sticky tape or small dabs of adhesive at the end of each strip, to make changes possible. This will allow you to ensure that the first line of a new paragraph does not begin at the end of a column, or a new page begin with the last line of a paragraph. Do not try to fix these problems by changing the line spacing of the affected sections. In some cases it may help to bring back a word or two to each previous line to shorten the text block, but the lines must never become so packed that clarity suffers. At the start of a new chapter line spaces may be added or deleted to manipulate the layout, but the change should be made for all chapters. It is acceptable to shorten or lengthen the text block by a line, but if this becomes necessary too frequently, it is better to redesign the entire layout.

The preliminary layout of the entire book, rather than of single pages, makes it possible to judge the effect of all elements in context. When all aspects of this version please you, start work on the final copy.

To make guidelines on your paper, stack six to ten sheets exactly even and put your layout sheet on top of the stack. Now mark the end points of the lines with an awl or a pin, punching vertically through the stack. Take care not to shift the sheets during this process. Afterwards draw fine lines with a pencil or with a scriber (see Figure 369), which will make later erasing unnecessary. Draw guidelines for the side margins. Careful work is crucial, because deviation from the original measurements will cause a change in the proportions of the letters.

Binding Handlettered Books[9]

Paper Structure

Any work more involved than a mere exercise and certainly anything that is meant to be bound later on requires some elementary knowledge of the properties of paper. If a paper has a smooth and a distinguishably rough side (the side that contacts the screen during papermaking), use the smooth side if possible.

Fold sheets of paper parallel to the grain direction, or the pages of the finished book will not turn easily. It is especially difficult to fold heavy papers or cardboard against the grain. Adhesive bindings also require folds in grain direction.

Here are several methods to determine the grain direction:

1. Visual inspection. You may be able to see how the majority of long fibers became aligned on the screen during production.

2. Tear a scrap of paper in both directions. The paper will tear more easily and evenly along the grain.

3. Bend a sheet of paper both ways. There will be less resistance when you bend it along the grain.

4. Run your thumbnail along a folded edge. The fold in the grain direction will remain flat, but the fold across the grain will become wavy or even wrinkle.

---

9. This section was written in cooperation with the bookbinder Renate Herfurth, Leipzig.

Folding Paper and Board

Remember to fold paper with the grain. Align the upper left with the upper right corner of the sheet, match the edges carefully, and hold the sheet steady with one hand while you push down on the fold with your thumbnail or a bone folder. Start in the middle and smooth up- and downward.

To create an accordion fold, determine the height and width of the finished product and cut your paper strip in the needed size. Fold the left edge of the strip over in the desired width, turn the strip around and repeat the procedure. If the strip is not long enough for the needed number of folds, leave an allowance of ½ inch (1 centimeter) after the last fold and glue the next section of paper strip to it. The first and last folds form the covers.

Folders, Covers, and Tubes

The easiest way to make a folder for a single or doubled sheet is to cut a piece of lightweight board to the height and twice the width of the sheet that is to be protected. Add an additional one-third of the width for a flap and 1/16 inch (1 millimeter) on all sides as overhang (Figure 396).

Another method of making a board cover is to fold in half twice a piece of board four times the finished size (Figure 397). Cover the board with paper (Figure 398). These folders may be used for single or folded sheets.

To secure a folded sheet in the cover use cording, a thin strip of leather, or better yet, of parchment, around the fold and tie the ends at the bottom of the folder. Slip the lettered page between folder and cording.

For particularly important projects, you may wish to consult a professional bookbinder and delegate this task to him or her.

Single sheets, like diplomas or certificates, can also be rolled up and secured with a leather thong, or the roll may be placed in a parchment or leather tube. A little-known procedure is to line the inside of a sheet of parchment with paper in a hue that matches both the parchment and the sheet with the text, and to attach the document to it along the upper edge. A leather loop serves as closure.

Binding Multiple Pages

Several methods can be used to bind a number of handlettered pages.

Simple Binding. Arrange no more than four folded sheets to make a signature: the number of sheets depends on the weight of the paper and on the total volume of the finished book.

More involved pieces merit the use of endpapers, which are pages that connect the book block to the cover. One part is attached to the inside of the cover, the other half is loose. Beyond its practical purpose, the endpaper also has aesthetic value. It can be made from the same paper as the rest of the book or from a decorative paper in a complementary or contrasting color. If you use endpapers, be sure to fold them with the textured side outward and the smooth sides facing each other.

The cover for a single signature can consist of two folded sheets of paper, one inside the other. Two or more signatures require a more complicated cover. Glue a folded page onto the first page of the first signature and onto the last page of the last signature. Apply adhesive along ¼-inch (5-millimeter) wide strip and parallel to the fold (Figure 399) and set it onto the first signature. Set it slightly away from the fold, to ensure that it will not be caught by the needle during sewing. Smooth the glued area with your hand.

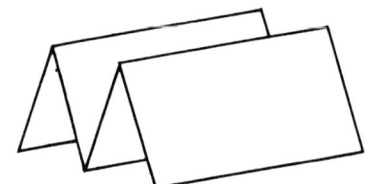

393 *Accordion fold.*

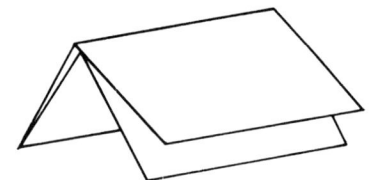

394 *Parallel fold.*

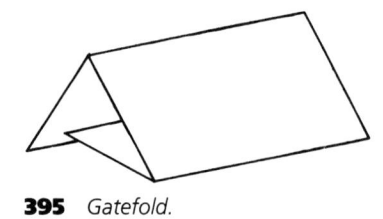

395 *Gatefold.*

Figure 396

Figure 397

Figure 398

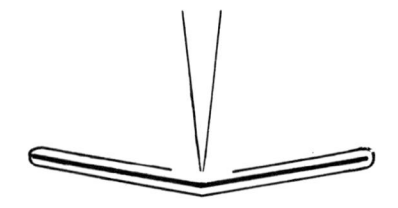

399

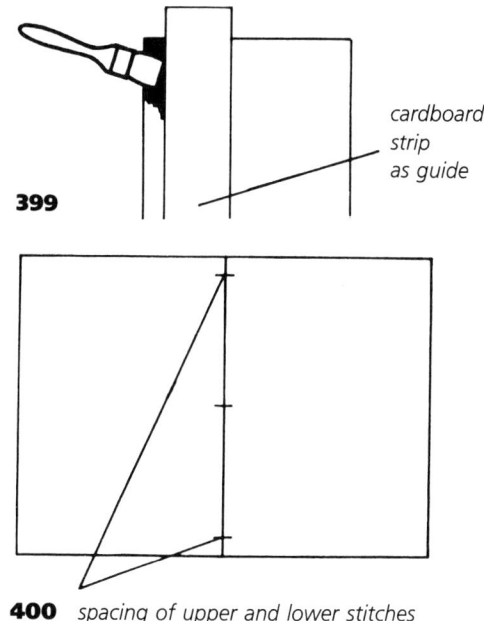

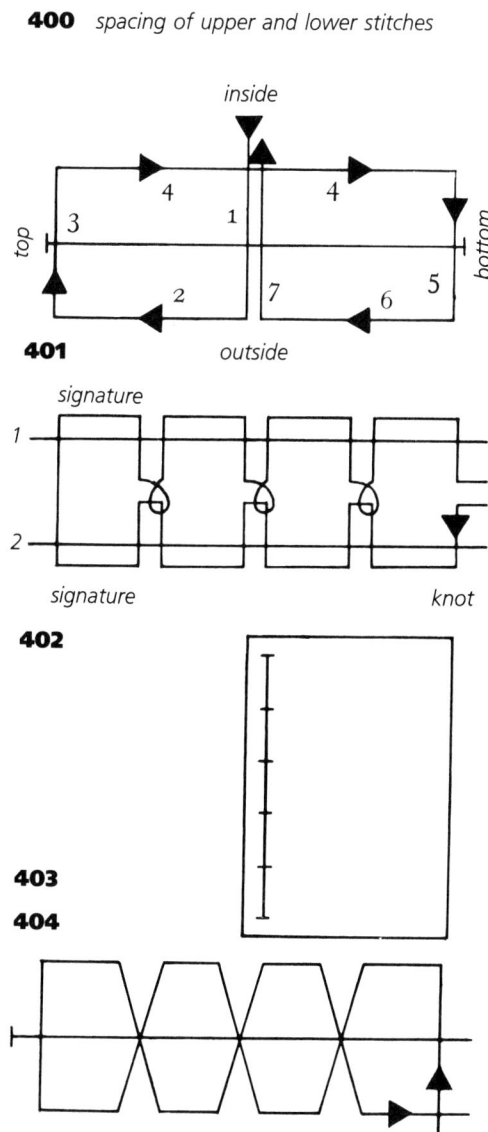

400 spacing of upper and lower stitches

401

402

403

404

To sew you will need a needle and sewing thread of natural or other colors; twine and narrow strips of leather or parchment are all suitable materials. Arrange the signature so that it lies parallel to the edge of the table and open it. Stitch from the inside out, starting in the middle of the sheet. Stitch back in at a distance of about 1 inch (2 centimeters) from the upper or lower edge. Repeat the process for the other side: push the needle out through the same center hole and back in near the other edge at the same distance as before. Tie the ends of the thread and cut them short. It looks good if the distance from the lower edge is kept slightly wider than the distance from the upper edge (Figure 400), or if the upper and lower stitch holes correspond with the edges of the text block. Figure 401 shows the path of the thread.

If you want to sew two signatures, put them on top of each other and arrange them as before. About 1 inch (2 centimeters) away from the lower edge start stitching from the outside in, continue out 1 or 2 inches (2 to 4 centimeters) further, stitch back in, and go on until you reach the other edge. It is best to mark the position of the holes with a pencil before sewing. Connect the two signatures by picking up the threads of the first one when you sew the second. Keep the threads taut, knot well at the end, and cut the threads short. Figure 402 shows the path of the thread.

It is difficult for the novice to bind more than two signatures. The techniques involved are hard to master and best left to the professional. The methods described earlier cannot provide enough stability for a substantial volume, which requires reinforcement tapes, a guillotine to trim the block, and other equipment, as well as dexterity and practice.

The Side-Sewn Book. Hold the block between thumbs and index fingers of both hands and knock it gently against a table top, upper side first, spine last. Arrange the block parallel to the edge of the table and punch holes with an awl at intervals of about 1 to 3 inches (3 to 8 centimeters) and ¼ inch (5 millimeters) from the spine. There should be an allowance of ½ inch (15 millimeters) from the upper and lower edges. Punch from the back side. Figure 404 shows the path of the thread.

A side-sewn book can also be held together by metal fasteners. These are easy to attach, but they may rust, and the block usually needs an additional cover to hide the unattractive fastenings.

In side-sewing the threads are on the outside of the covers, and sometimes encircle the spine as well (Figure 405). If strips of parchment or bleached threads are used, the effect can be quite decorative.

Side-sewing can be used for single sheets or for folded sheets with the fold placed towards the spine of the book. The first and the last page double as endpapers. However, additional stability as well as elegance are gained if the folds are placed at the outside, or fore-edge, of the book rather than at the binding edge (Figure 406). This page arrangement, combined with the side-sewing technique, is known as Chinese or Japanese binding. It has several benefits. If the paper is not opaque, you avoid show-through because you are writing only on one side of the sheet. Thus, if you make an error, you will only have to rewrite two pages, rather than four.

Side-sewn books can be glued into an additional cover, but it is more common to attach the covers to the block by sewing. Use thin cardboard covered with heavy paper in a nice color (Figure 407).

Adhesive Binding. Adhesive-bound books are held together not by sewing but by a special glue. Adhesive binding is commonly used for soft-cover books, but is equally useful for larger handwritten books, because the sheets can be arranged with the fold towards the outside, as described above. The pages open more freely in this style, but the book should be assembled by a bookbinder, and it will not be as durable as a conventionally bound book.

Covers for One or Two Signatures. A board cover is easily fashioned. You must crease the board parallel to the grain direction in the middle of the board and sew or glue the signature into it. If you are gluing the signature, you must score two more lines parallel to the middle fold of the board, one on each side, at a distance of ⅛ inch (5 millimeters) from the fold, so the book will open easily (Figure 408). Paper can be folded around the board cover and held in place by means of flaps that reach almost back to the fold (Figure 409). If you use this method, which is known as English binding, it is not necessary to attach the first and last pages to the cover—just put them under the flaps. The wrapper can reach as far back as the inner edge of the cover and can be made of printed, marbled, or otherwise decorated paper. It carries the title either in written form or on a label.

When you score board for a booklet of two signatures, you must take the width of the spine into account (Figure 410). Apply adhesive to the cover, insert the signatures, and weight the book between two wooden boards until the glue is dry. Depending on the size and thickness of the block, you may need to attach the endpapers to the cover with adhesive, or you may just tuck them under the paper wrapper, as described above. Usually the latter is sufficient.

Should you desire a traditional hardcover binding, consult a hand bookbinder. It remains your responsibility to make suggestions for the design and to choose the materials.

We have covered only a limited amount of information about paper, covers, and techniques of binding. The graphic artist who wishes to know more can consult Franz Zeier's *Books, Boxes, and Portfolios* (New York: Design Press, 1990) and other sources.

Lettering in Applied Graphic Art

Logotypes

A logotype is an expression of the essential substance of a particular company, institution, or organization, of an idea, a special occasion, or a product. Logos are visual signs, and their form comes from the object they depict or from related associations. Geometric or natural forms can inspire logotype designs—for example, the environment could be symbolized by a stylized leaf. Other sources are scientific symbols, heraldic forms, or visual representations of the word in question. The locality of a subject might be important enough to be pictured. Purely naturalistic forms, however, are rarely effective.

A logo can also be developed from a company's initials or the name of the product. Logos that are made up of letters, monograms, and lettering of any kind demand the same treatment as pictorial symbols. There must be contrast and tension. Interior space and spaces between shapes carry as much weight as the shapes themselves, and all the graphic elements have to form a unit. A logo should be more than a conglomeration of unrelated elements that are held

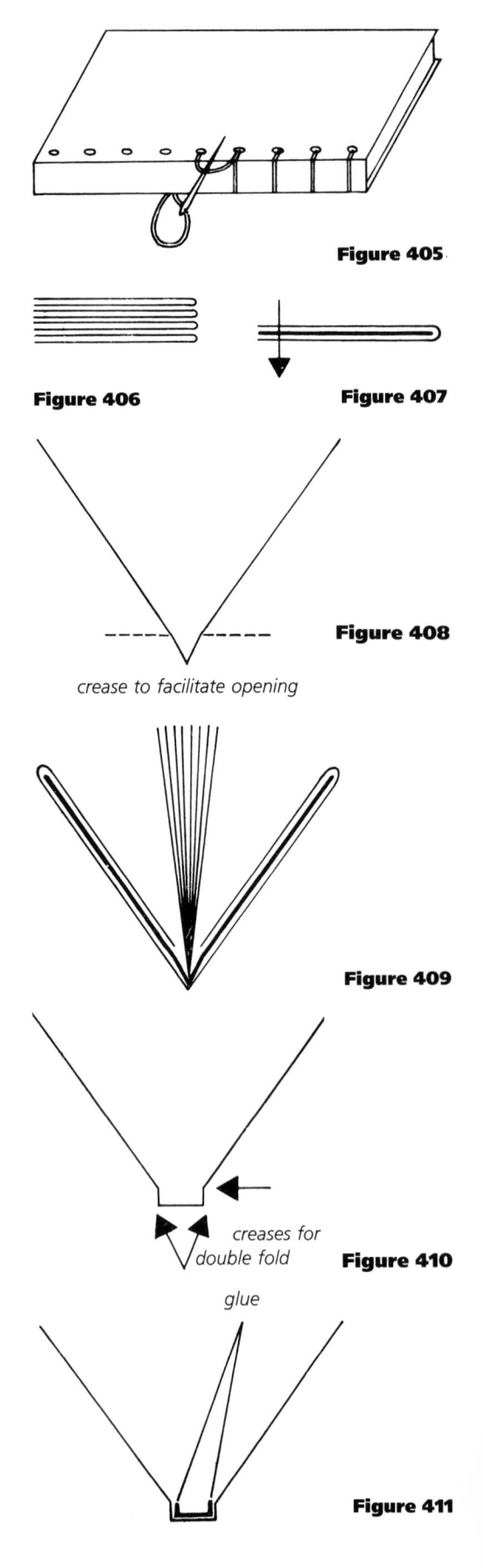

Figure 405

Figure 406   **Figure 407**

Figure 408

crease to facilitate opening

Figure 409

creases for double fold **Figure 410**

glue

Figure 411

together by a border. The examples in Chapter 5 illustrate this principle.

It is, of course, possible to combine letters and pictorial elements in a single logo. Certain letters provoke associations such as a feeling of lightness or weight. Use these associations as well as any other emotional messages the forms may carry.

The following principles apply for all logos, pictorial or based on letters:

A logo must be easily recognizable; it has to be simple and memorable.

The purpose of the logo should influence its form.

Most graphic forms of advertisement are based on or include the logo of the subject; the logo usually appears on letterheads, brochures, labels, packaging, and delivery vans. It may be necessary to render it in varied materials, such as cardboard, plaster, glass, metal, fabric, or even neon. The technical requirements and restrictions of work in any of these materials must be taken into consideration from the earliest stages of design, since it is obvious that printing, embossing, punching, casting, or weaving require distinctly different approaches. Variations may be necessary if the same logo is to be executed in techniques as different as engraving in steel or modeling in plaster. It is rare, however, that one design has to fit such diverse requirements; more commonly it is enough to satisfy the following requirements.

All details should still be visible if the logo is reduced to ⅛ inch (5 millimeters). Unlimited enlargement should be possible, though a variation of stroke thickness might be necessary for very large versions.

The logo has to be reproducible in black and white and positive or negative, and it is useful if a representation in several colors is programmed into the design, but it is rarely feasible to concentrate on color exclusively. It has to stand on its own as well as fit into a frame. Consider the possibility of relief or freestanding sculpture.

Logos are protected by law. A new design must be original and may not create associations with already existing ones. The simpler the design, the harder to avoid this problem.

Logos are subject to fashion. The taste of the public changes in the field of graphic art almost as quickly as when it comes to hem lengths. Since the logo is an essential element of all graphic design pertaining to a product, it should be changed only if absolutely necessary, and then only gradually, especially if old and established products are concerned.

Some logos are developed from the name of a product or a company. The letterforms should be chosen in relation to the particular product or company; the letters must form a word that stands on its own as a composition, and the word must stand out among other text elements. Again, the design must be easy to remember.

To ensure that the word can be read easily, the letterforms must never be modified beyond their basic characteristics. Figures 412 and 413 show examples of design that sacrificed legibility to the visual image in a misguided attempt to be original. Unfortunately, many similar examples could be cited.

Logos may be executed with pen or brush as well as with type, but the field of application is more limited, because calligraphic forms cannot be transferred so easily to other media.

To get started designing a logo, make a number of preliminary sketches. If one of these looks promising, play with it in a small format of about 2 inches (5 to 6 centimeters) in black and white. Figure 414 is an example. The topic for this particular exercise was to inscribe a cap-

412 *Poor graphic design.*

413 *Poor graphic design.*

ital A into a circle. Make sure that you try all the possibilities you can imagine. New ideas or at least new points of view will surface during the process. Choose the best sketches and draw them more carefully at a larger scale—about 7 inches (15 to 20 centimeters) high. Now check to see if a photographic reduction to ⅛ inch (5 millimeters) still renders clear details.

Lettering in a Circle

Sometimes you may want a circular arrangement of letters. To balance the amount of text with the diameter of the circle, it may be preferable to arrange short texts in two semicircles. Capitals usually work best, because they create a more pronounced ribbon effect.

Figures 415 through 418 illustrate the process of arranging the letters in circular fashion. Start with rough sketches to establish the relationship between letter size and diameter of the circle. Include a border in the design. When you have decided on one design, multiply the inner diameter of the circle by 3.14 (pi) to determine the length of the letter strip (Figure 416). On this strip make a sketch and then a more detailed drawing of the lettering. Transfer it onto the middle line of the ring (Figure 417). Dividing the text into an upper and lower section will make it easier to read.

Posters

Most posters combine pictures with words. The lettering carries the specific message, but it is an integral design element and visually subordinate to the picture. Most graphic artists lack the training to design letters, and prefer to incorporate copies of established models into their work: their decisions are thereby limited to choosing type and determining its size and placement on the poster.

Alternatively, the lettering could be developed from the style of the image. Here the words would not act as a contrasting element, but rather as part of an integrated design. (See for example the poster in Figure 498, page 225, which did not require a knowledge of formal letter construction.)

In some posters the letters are the sole transmitter of meaning—that is, the text *is* the visual image. Handled in masterly fashion, the effect can be quite impressive. Some topics do not lend themselves to graphic depiction and are actually better served by such treatment. Typographic, handlettered, or even three-dimensional letters can be used and manipulated, and many combinations are possible. The rich diversity of content must be expressed through various representations of letters. A word or a group of lines simply formed by adding one letter to the other can surprise, and possess a decidedly monumental look.

Posters aim to transmit ideas, which may have cultural, educational, political, or other meaning. Unless the poster is displayed indoors only, it has to compete—on billboards and walls, for example—with a host of other materials for the attention of the viewer. Since it is

Figure 415

Figure 416

Figure 417

Figure 418

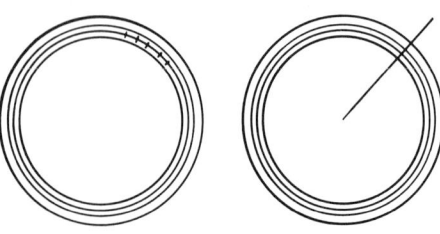

Figure 414

impossible to predict what the eventual surroundings of a poster will be, the designer is obliged to search for ever novel and interesting layouts. It should be noted here that it is not necessarily the most extroverted and gaudy image that draws the most attention, but sometimes the simple and serene.

A poster designed to be hung outdoors has to be visible for a distance of about 10 to 15 yards (10 to 15 meters) to attract attention and interest. Its drawing point might be an expressive quality, a feeling of movement up and down or back and forth, the emotional content of its colors, its intensity, or even its aggressiveness. Whatever the design, it must create quick associations in the viewer to involve him or her emotionally. Big closed forms with interesting details in the interior serve the purpose best, because they are visible from afar and still provide interest for the nearby observer.

A poster persuades by making its main message clear, and it is obvious that the words here are as important as the formal design. The central idea must be accessible quickly and without great effort, but this does not in any way preclude the inclusion of secondary messages that linger in the mind of the viewer for some time or become obvious only after repeated encounters.

To get started designing a poster: jot down ideas with pencil or colored pencils, then clarify important aspects such as size and weight relationships and color in small-scale sketches. Compare sketches of various elements in different sizes, stroke widths, and colors, and try out various arrangement of the cut-out pieces until the result is pleasing. Check to be sure that effects planned in the sketch format work when enlarged to poster size.

If the poster contains a lot of text, break it into blocks. Even in a poster made up entirely of text, certain sections can be planned as preworked units, to be replaced as necessary. For example, alternate information on the location of an event can be set on an overlay, to be placed in the appropriate position by the printer.

Packaging and Labels

A product wrapper simultaneously protects, advertises, and informs purchasers about content and quality. Buyers usually transfer the impressions they get from the package to the product itself. It is therefore obvious that the packaging should produce positive associations. Color, quality of material, and design are the key features to which the consumer responds.

The packaging designer needs not only technical skills, but also a certain knowledge of psychology. The designer has to be able to predict buyers' reactions, as well as convey information about the product. A can of cocoa must not resemble a paint can; toothpaste and skin-lotion containers have to be distinctly different.

The most important function of packaging is to characterize the product through visual elements—images, decoration, and lettering—all of which have to be integrated in the completed design. Type, photographs, and illustration can be combined. If the main element of design is the lettering itself, the ornamental values of the letters have to be explored for their aesthetic impact, but the wish to create something original must never interfere with the legibility of the text.

It may be necessary to include some or all of the following data on a package: contents, name of the manufacturer, price, weight or piece count, logo or trademark, governmental notices, ingredients, and assembly instructions. This information should be clear and easy to understand. Some of it, such as assembly instructions, recipes, or a list of parts, can also be given on the inside of flaps, or on a separate sheet of paper or brochure inserted in the package.

The graphic elements chosen must serve to distinguish the product from others to avoid confusion between similar products, but possess a recognizable style that is common to all products of the particular company. This can be accomplished by using the same lettering style on all printed matter, including the company logo and its signage. A unifying color or combination of colors can serve the same purpose. Entire families of product designs can be created in this way. Some companies or products are already well known and associated with a particular image or tradition. It is wise to continue such traditions if they are at all acceptable from the designer's point of view.

It is essential to consider the available materials for packaging and the ways in which they can be manipulated. Paper, cardboard, metal foils, vinyl, glass, wood, and sheet metal all have distinctly different characteristics, and react differently to print or other means of decoration. Other important considerations for package design include: economic factors and special requirements for handling, storage, and shipping, but these cannot be discussed in the limited space of this book.

Packaging is three-dimensional, and it is imperative to make full-size models to account for possible technical problems. Consider different views and angles of the finished product and position graphic elements accordingly. Boxes should present the intended view from at least two opposite sides, and the customer must be able to see the full name of the product even on round packages, where the area of design should not exceed one-third of the circumference. Think about what the package will look like when it is opened: it may be desirable to extend the design to the interior surfaces. Make several sketches of each element that you plan to include on the package, using different colors and sizes, and arrange the cut-out sketches on the model to determine what combination works best.

Labels identify products. They can be glued or tied on, or, in the case of textiles, sewn in. Labels may include any and all of the same information that appears on packaging: contents, name of manufacturer, price, and so forth. Again, this information and the advertising message is communicated through visual images. Keep the same objectives in mind when you design packaging and labels; the only difference is that packages are three-dimensional, while labels are flat, though in some cases a label may be placed on a curved surface, which could distort the image. As with packaging, a model can provide the testing ground for your design.

Book Jackets[10]

To protect the book it covers is only one reason books have jackets: the main function of a jacket or cover is advertisement. The purpose of a book jacket or cover, and its effectiveness, is very similar to that of a poster. In a store window or on a shelf, a book is in direct competition with other books, and the visual images that are created resemble those we receive from a row of posters. The effect of a bold design will be enhanced by a neighboring subtle and discreet look, and vice versa. Undoubtedly, contemporary posters and book-jacket design influence each other.

A book jacket or cover should not only be attractive and tasteful, but should also mirror the content and spirit of the book. It is of course impossible to give concrete advice on graphic means that would befit specific situations, since the variety of possible topics, themes, and values is endless. A list of feasible treatments includes graphic illustration with wood- or linoleum cuts, pen or charcoal drawing, painting, photography or photomontage. As in the case of the poster, the specific information is transmitted by the written word, which has to be included in the design and may even take a subordinate position to the graphics.

Ornamentation is appropriate for historical, national, or modern literature, where proper connections can be made to the content of the book. Here, too, the specific information is carried by the written word.

Almost any topic can be adequately represented through lettering alone. Letters can be drawn, typeset, or manipulated photographically, and there is an almost unlimited variety of combinations. The effect can be bold or reserved, but no amount of originality justifies lack of legibility.

A jacket or cover has to suit the book it is designed for. Colors, layout, and type have to be unified, but many variations are possible. For instance, you might use the text type for the advertising copy on the jacket to establish a connection between inside and out. Select type for books and jackets according to the guidelines given in the preceding chapters. Contrasts will probably yield the most eye-catching results. Most design efforts will be concentrated on the front of the jacket or cover, but the back can certainly be included, as long as both sides can be viewed as independent units.

There are no strict rules for the placement of type on book jackets or covers. The name of the author usually appears on the front; it can also be given on the back. The text on the spine can read vertically or horizontally, depending on the available space or the specific design. If a bellyband (a paper strip with a commercial message on it) is planned, it has to be included in the design, or it might cover important information on the front.

The colors of the jacket can be more distinctive than those of the binding, but you should relate them to each other as well as to the content of the book itself. Technical consideration of printing and finishing could limit the choice of colors. If the binding of a book by itself is attractive enough for advertising purposes, a clear plastic jacket could be used in place of the traditional paper dustjacket. In most cases, however, the binding is more subdued, and a jacket is added for advertising purposes as well as to protect the case.

Lettering for Exhibitions

The lettering that is used for exhibitions must inform viewers about the articles on display, their history, purpose, properties, and economic uses, and it must do all this in a complete and instructive way. To accomplish this task it is useful to create a kind of script, containing all necessary information, as well as relevant facts about the architecture of the location, the lighting conditions, the traffic flow, and the distance of the lettering from the viewers. A cardboard model in a scale of 1:20 is an invaluable

---

10. This section follows the outline of Albert Kapr in his work *Buchgestaltung* (Book design), Dresden: VEB Verlag der Kunst, 1963.

aid. The character and the effect of the lettering must be subordinate to the requirements of the total graphic concept of the exhibition.

The best basic type styles are those that are neutral and have unobtrusive detail, such as sans serifs like Helvetica and Univers, condensed sans serifs, Egyptian faces, Garamond, and Bodoni.

Ideally, all lettering design for an entire exhibition, or at least for each display unit, will be based on the same family of typefaces, to create a coherent effect. This would not preclude the use of visual focal points or highlighting techniques to differentiate objects from each other. Superficial or forced presentation and pretentiousness, however, are never justified.

Many modern developments have made preparation of type for exhibitions easy. Rubber stamps and templates or stencils, cutout cardboard, foam-core, and wood letters, and various forms of transfer type are available, as well as type produced on the personal computer. If you use prefabricated letter materials, make sure that they are of high typographic quality, and assemble them in accordance with the principles of good lettering. Too often, particularly where long texts are involved, letters are stamped or laid out too close to each other, the line spacing is too wide or too narrow, and the entire text is forced into a text block to achieve justification, without any concern for sense or aesthetics. Even the basic rule that round letters should exceed the boundaries of the x-height slightly is all too often ignored when prefabricated letters are used.

Technical innovations have made lettering easier, but this does not mean that the field of typography and lettering can now be left to amateurs. The new tools and procedures should merely facilitate attention to detail.

Rubber Stamps

Stamping can only be successful if the rubber stamps are assembled with utmost precision. When you glue letters to the wooden block, position them so that the left edge of the letter is flush with the edge of the block. Make sure that the base line of all letters is at the same distance from the bottom edge of the block, which means that you have to leave room for descenders even in letters that do not have any. Use only the most reliable adhesives.

You can stamp on any material with a smooth surface, but thin materials like paper or fabric must be supported by something hard and smooth, like a sheet of glass. Choose colors that cover well, are stable, lightproof, and weather-resistant if necessary. They should also have a short drying time. Inks made for offset printing work well on absorbent grounds as well as on glass, sheet metal, plastic, and aluminum foils. Drying time and printing surface determine the choice of medium.[11] For example, on a ground of emulsion-based paint, offset printing ink dries in two to three hours, while silkscreen colors take at least seven hours, longer for certain hues. Offset inks on a ground of latex paint need three to four hours to dry, silkscreen colors seven to eight hours, and ink for linoleum-cuts, ten to twelve hours. On photoprint paper, offset ink dries in four hours, silkscreen colors in eight to ten hours, and linoleum-cut ink in twelve to fifteen hours. On paper with a high wood content, offset ink dries in two to three hours, glossy silkscreen color in seven to eight hours, and linoleum-cut ink in twelve to fifteen hours. On glossy papers, it takes four to five hours for offset ink to dry. On glass,

---

11. All the information on drying times in this section is drawn from texts of the Betriebsberufsschule Handel (Business School), Halle.

offset inks take two hours to dry, matte silkscreen color ten hours, and linoleum-cut colors fifteen to eighteen hours.

You can also print with offset, silkscreen, or linoleum-cut colors on fabric. Use a heavy application of color, because the ground is very absorbent.

Certain inks cannot be used for stamping because they would react with the rubber, and oil-based paints, poster paint, and tempera produce uneven applications and fuzzy edges.

Use ink made for rubber stamps straight from a stamp pad. If you are using offset printing ink, spread a thin and even layer of it on a sheet of glass with a rubber roller and ink your rubber stamp from that surface. Add thinner or drier to make the ink more elastic and speed up drying time.

If you allow paint or ink to dry on rubber stamps, they quickly become unusable. Clean them immediately after each use with the appropriate solvent for the medium used, but do not use gasoline or alcohol, because over time these substances turn the rubber hard and brittle. Treat rubber rollers the same way.

Use a wooden or metal guide to align the stamps. It should be about ½ inch (15 to 20 millimeters) thick, 2 inches (4 to 5 centimeters) wide, and about 3 feet (50 to 100 centimeters) long, but it is even better to have several of them in different lengths. Instead of marking the entire line of type, draw small dots at the end of each line. If you work on glass, put a lined sheet of paper underneath. Now place the guide parallel to the marked lines at the right distance so that the baseline of the stamped letters will fall on the guide line. Hold the guide in place with your free hand while you stamp. If you glue thin rubber strips under both ends of the guide, it will stay in place well.

195

Set the stamp down with even pressure, so that edge of the wood block that is closest to you rests tightly against the guide. If the left edge of the letter is identical with the left edge of the block, it should not be difficult to achieve the right spacing. If you find blank spots within the printed letter, fill them in with a brush after the ink has dried.

Stencils

Cut your letters from stencilling paper, which is varnished drawing paper, or from see-through plastic sheets if you plan to use them often. Keep the distance between the baseline of the letters and the bottom edge of the stencil equal for all letters to make it easier to line them up. Cut the side edges in a true right angle to the bottom edge. Mark the line spaces at the side of the paper, draw the lines, and put the stencils in place. It is easy to find the right letter spacing if the stencils are clear material. Hold the stencil tightly and stipple the paint on with a stencil brush or spray it with an airbrush. If you wish, you can later fill in the areas within the stencilled letters, but it is not necessary. Use thick paint that adheres only to the outer ends of the bristles.

Computer-Generated Lettering

Neither conventionally typed copy nor ordinary printer output from a word-processor will yield satisfying results when enlarged. Laser proofs are better, but still not intended for enlargement. All of these methods of producing lettering will yield graininess and fuzzy (jagged) outlines. For good quality type, with crisp contours, use professional imaging (typesetting) equipment such as the Linotronic 300 or Compugraphic 8600.

Drawing Large Letters

To achieve a precise rendition of large-scale letters drawn by hand, put the entire alphabet on tracing paper and follow the outlines with a tracing wheel. Transfer the letters either with the help of carbon paper or a pounce bag. Later the dots can be connected and the letters filled in. Another method is to cut the letters out and trace their outline.

Transfer Type

Several companies produce a wide variety of transfer type — alphabets and other type elements as well as symbols, grids, tone, and areas of color — printed on a thin film with a transparent protective covering. These letters are pressure-sensitive and can be rubbed onto a smooth surface, primarily paper, after which the protective covering is removed. The resulting type has precisely defined outlines and is ready for reproduction. Instructions for use are included with the products.

Graphic Arts Tools and Procedures

Roughs and Finished Artwork

If you accept a commission for work that will be printed, get all the relevant information about the format, number of colors, and the printing technique as soon as possible, since the resulting restrictions will influence the design process from the very beginning.

You will have to make several sketches — most clients expect at least three — but great detail is not necessary at this stage. Finished artwork is made for reproduction. On it indicate the trim marks, so it is clear where the image is to be posi-

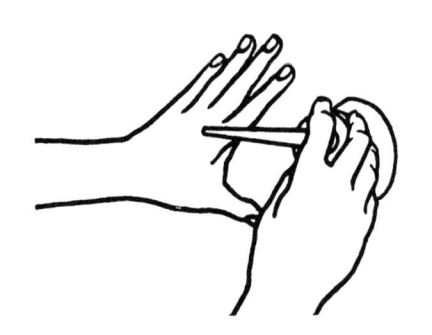

419 *How to hold a brush for stencilling.*

tioned. If lines or areas of color or other elements are to bleed—that is, extend up to the edge of the poster, book jacket, or label—draw them slightly beyond the trim on your final artwork to avoid problems should the trimming not be exact. The color on the final art has to be completely opaque and should not have any shiny or transparent spots, which could complicate the reproduction process and require retouching. Drawing ink is therefore not a good choice: tempera or gouache colors are better. Use real Chinese ink in bar form for small letters and add a small amount of gouache color to avoid transparency and glare. Too-liberal use of opaque white for corrections will form raised edges when it dries and cause shadows during photographic reproduction, so the contours of letters may appear fuzzy. Repairing such defects is time-consuming and expensive at the printing stage, and a loss of quality can not always be avoided.

Execute your design in black and white even if it is to be printed in color on white or any other color background. If your design is to be in gray created by the use of a printer's screen, indicate the percentage screen to be used. If you combine lettering with other elements in full tone or other colors, you have to be sure that the letters will "read," especially if filters are used. For combined line and tone projects, make a separate rendering of the text in black and white on an overlay, and be sure to include registration marks on both parts.

Elements such as color gradations, photographs, and other toned items (for instance, the mark of a pencil or pastel crayon) are usually reproduced with a halftone screen. If the original is multicolored and combined with lettering, again two-part art in black and white needs to be prepared. Use positive letters on multicolored backgrounds, because the dots of the halftone screen produce imprecise outlines for the letter shapes and a superimposition of several screens would aggravate the problem considerably. Another difficulty of using small, negative (reversed-out) type or hairlines is caused by the fact that large areas of halftone require a lot of ink, which may fill in thin spaces that should remain open. Such designs are best done in offset printing.

Lettering for lithography needs substantial elements: fat-face types are not suitable, because their thin lines and diagonals tend to look blurry.

Include exact information about color with your finished artwork. Color samples should be at least ½ inch (15 millimeters) square. Build your design on standardized colors whenever possible.

A host of problems can confront a graphic artist in daily practice. There are many books that deal with preparing art for reproduction and other practical matters: among those that are recommended are *Studio Tips* by Bill Gray (New York: Prentice Hall Press, 1976) and *Preparing Art for Printing* by Bernard Stone and Arthur Eckstein (New York: Van Nostrand Reinhold, 1983).

Sketching Type

Typesetting requires careful preparation. Sketches or rough layouts should represent the character of the chosen type as accurately as possible; the dimensions and color effects have to be considered carefully. Type specimen books are indispensable; many good ones can be purchased, and many typesetters will supply their own. Unpleasant surprises come from sloppy type sketches. The more precise and professional the roughs supplied, the more fruitful will the cooperation between graphic artist and typesetter be.

Rough sketches can be done in pencil. Use hard ones for small type, medium-hard pencils for mid-sized letters, and soft pencils for large letters. Sometimes the use of pen or brush and ink is necessary to represent the gray values of different types adequately. Black ballpoint pens are practical for mid-sized type, with red or blue pens for emphasized letters. Six- to twelve-point type can be drawn with pencils of appropriate widths in the required gray value.

Papers with treated surfaces are usually impractical for sketching because they show eraser marks, but a smooth surface is essential. You will also need a drawing board or similar support, a right angle, compasses to transfer measurements, and, last but not least a type ruler such as a Haberule, since typographic sizes are the basis of all measurements.

Get a feeling for the character and flow of the chosen type by copying it onto tracing paper. Make pencil lines for the baseline and x-height line, and draw ascenders and descenders freehand, preserving the correct proportions. The special aspect of this sketching technique is that line is added to line until the right stroke width is achieved. It would not be appropriate to draw the outlines of the letters first and to fill them in later, because this would create misleading effects. Corrections would abound, and the flow of the copy would be difficult to achieve.

Lettering on Photographs

Superimpose letters on photographs only if they constitute a design element, and place them in areas of the picture that have as little detail as possible, or the text will be hard to read. Ideally, the position of the letters will have been planned before the picture is taken.

If you are dealing with final artwork and your original is a photograph, do not draw directly on it. The smooth surface makes lettering difficult, and the pencil lines will leave marks on the finish. Mistakes cannot be fixed with opaque white, or, in the case of negative letters, with black, without disappointing results. Retouching errors could only be corrected by washing the photo, and it would be impossible to use it a second time. It is a better procedure to make a separate proof of the text and let the printer superimpose it phototechnically on the picture.

Phototypesetting

Developed in the 1950s, phototypesetting rapidly supplanted the various metal typesetting methods over the following twenty years. Phototype, produced from one, or at most two, master pieces of art (type) could be enlarged to any size and in any configuration both quickly and economically. The use of different lenses in photo reproduction made it possible to change the height and width proportions, slant letters (oblique) and distort them.

The maturation of phototypesetting in the early 1960s made it possible to overcome many of the major limitations of metal type, whether foundry or machine. Phototypesetting removed the physical aspect of type. No longer was the type body immutable (or nearly so). Letters could be set close to one another — both horizontally and vertically — as one wanted. They could even be overlapped. The shoulder of metal type and the set width of individual characters could be violated. This meant that letterspacing and kerning could be more easily perfected. *We*, *Ty*, *Yo* and other troublesome letter combinations could now be fixed without the use of saw or file. Letters such as *f* or *y* could be properly kerned or fitted together, allowing designers to draw them without any compromises created by body width or duplexing (forcing italic letters to the same width as accompanying roman). Since phototype did away with metal letters, it also did away with metal spacing material. Leading became an anachronistic term. Line spacing, or linefeed, could be adjusted in fractional point increments. It could also be done "negatively," making the space between lines less than what would have been achieved if the type was set solid (with no leading). This was especially useful where descenders and ascenders were infrequent or, as in much advertising, copy was brief.

In freeing type of its physical constraints phototypesetting inevitably made it lighter and less burdensome. Instead of storing enormously heavy cases of foundry type, typesetters had film strips, discs, or glass grids. The setting of large type, especially above 96 points, was now a simple matter of photographic enlargement. Weight was not an impediment. Similarly, typefaces could be created at any font size desired, not just the traditional incremental sizes of 6 through 14 points, 16, 18, 24, 30, 36, etc. Type could be made to fit a specific layout rather than vice versa.

Phototypesetting made type malleable. Individual characters could be manipulated at will. They could be stretched, expanded, condensed — even distorted — all by using a variety of camera lenses. Often typefaces were slanted to create companion "italic" faces where none existed before. And type could be slanted backwards as well as forwards.

Additionally, phototypesetting made it possible to combine multiple typefaces in a single line or word. A common

420 *Experimental design based on the letter O, by Werner Pfeiffer. Reproduced by permission of GRAPHIS.*

421 *Experimental designs based on the letters Z and C, by Werner Pfeiffer. Reproduced by permission of GRAPHIS.*

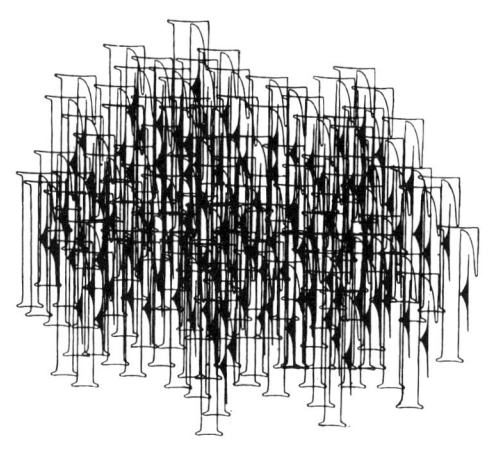

422 *Experimental design based on the letter F, by Werner Pfeiffer. Reproduced by permission of GRAPHIS.*

baseline was easily achievable.

The 1960s witnessed an explosion of new typefaces and a burst of typographic creativity and experimentation as a result of the change in technology. Since type designers no longer had to concern themselves with the limitations of punchcutting, engraving, and casting, they could create typefaces with more flamboyance and eccentricity. The camera allowed any set of drawn letters to become a typeface. And it made it possible to revive many of the older designs of the past, especially those of the frequently maligned nineteenth century. Much of the style of typography in the United States in the 1960s and 1970s can be attributed to the influence of phototype. But like all technology, phototypesetting has its disadvantages and drawbacks. Its flexibility led, in the 1970s, to overly tight letterspacing, not to mention frequent overlapping of letters. The ability to scale innumerable sizes of type from one, or at most two, masters has led to compromises in the visual quality of type when enlarged. Serifs and hairlines or thin strokes become ungainly. The minute adjustments made at each size in metal type for optical reasons have been lost. The general problems of photography—proper focusing, fresh chemicals, dimensionally stable paper or substrate, accurate exposure time, etc.—are all present in phototypesetting. One problem is halation which causes light traps or clogging where acute strokes meet in a letter. This is the photographic equivalent of the ink traps found in letterpress printing. Another problem is the soft edges and rounded corners that occur when phototype is greatly enlarged.

Computer typesetting came of age in the mid-1980s with the introduction of the Apple Macintosh personal computer, Adobe's PostScript page description language, and imagesetters such as Mergenthaler's Linotronic 300. Computer typesetting creates letters either through bitmapping (also known as rasterization) or outlining. Bitmapping places a grid over a drawn character and tries to approximate its shape. The finer the grid, the more accurately the character is reproduced. Outlines trace the edge of a drawn character, frequently through the use of complex mathematical equations known as Bezier curves, and then fill the enclosed space. The result is usually a more accurate rendition than with bitmapping, but more computer power is required. The fineness of the grid or mesh determines how subtle a bitmapped character will be, while the number and placement of control points for the Bezier curves influences the quality of outline characters.

Computer, or digital, typesetting, is an extension of phototypesetting, with similar advantages and disadvantages. Type can be scaled, condensed, expanded, distorted, and slanted both forwards and backwards; letters and lines can be set as close or far apart as wished, set along contours and curves, and so on. When type is created by outline (and the thickness of the outline can be adjusted), filled in solidly or with an infinite variety of patterns. And, of course, these effects can be combined. Such tricks turn letters into graphic images rather than type.

As with all typesetting technology, computer type is device-dependent. That is, the equipment that the image is printed on, even more than the equipment that it is created on, will determine its level of quality. Type may be printed on a dot matrix printer or a low-resolution laser printer (300 dpi or dots per inch) or on a high-resolution imagesetter (2540 dpi). In the former instance the type will appear jagged or coarse; the letter and wordspacing will be erratic; and letters will be heavier than expected. High-resolution devices solve these problems.

The flexibility and malleability of phototype has been extended by computer type. But the problems of good letter- and word spacing still remain. Computer memory and power is often an impediment to creating all of the necessary refinements for the myriad letter combinations found in languages that use the roman alphabet. The shoulders, set widths, duplexing, unit widths, and other type body parameters built into earlier generations of typesetting equipment may have created limitations and compromises, but they set overall standards that were easy to maintain.

An offshoot of the shifting nature of computer type is the ease with which it can be created, edited (as distinct from manipulated), and distributed. Anyone with a computer and one of several typographic software packages such as Altsys's Fontographer, Ikarus-M, or Letraset's Fontstudio, can change existing typefaces to suit their preferences and prejudices. They can also create their own typefaces. The type designer and manufacturer, once united in the person of the punchcutter, are reunited. Editing and creating can be done with both bitmapped and outline fonts, though the latter are more flexible. It can even be done using non-typographic software such as Adobe's Illustrator 88 or Aldus Freehand, since type is now a graphic element.

Special Techniques

Woodcuts and Wood Engravings

Both woodcuts and wood engraving are relief printing processes, which means that the image to be printed is left on the surface of the wood, while all other parts are cut away. The medium is a slab of soft plank- or side-grain wood — that is, wood cut along the grain — for woodcuts. Endgrain wood is used for engravings; the fibers are cut across the grain.

Woodcuts can be used to reproduce lettering, but they are primarily a means of artistic expression, since many other superior techniques have replaced the woodcut for the mere printing of type. From our point of view, the inspiration is usually a drawn or printed letter, but the woodcut should go beyond mere imitation in the use of materials and tools. The cutting process changes the character of the letters and makes them look rougher and angular. Attempts to recreate the pen stroke in detail interfere with the particular expressiveness of the medium. The initial design should anticipate the character of the finished piece.

Highlights in the history of woodcut lettering are the stamps of the Far East and Chinese block books, European block books of the second half of the fifteenth century, the woodcuts of the Expressionists, and finally the works of Rudolf Koch and the Stuttgart school under Ernst Schneidler. Any designer who wishes to master the intricacies of woodcutting is well advised to study these examples of the technique.

The ultimate effect of a woodcut depends largely on the materials used.[12] Whether the wood is hard or soft, fine- or rough-grained, the marks left by the cutting tool, the finish and surface — absorbent or polished, smooth or porous — of the paper, and the pressure from the printing press or the artist's hand determine the outcome.

There are two kinds of woodcuts — the positive and the negative; the latter is easier to handle and more satisfying

---

12. The following information on woodcuts and endgrain carving is excerpted from the work of Johannes Lebek, *Holzschnittfibel*. Dresden: Verlag der Kunst, 1962.

423 *Woodcut of the artist's name, by Karl Schmidt-Rottluff, 1920.*

424 *Woodcut by HAP Grieshaber, 1938.*

for the beginner to work with.

Several types of wood can be used for woodcuts: linden or pear and alder, even though it splinters easily, and willow, poplar, birch, and maple. Even hard woods like apple, cherry, and white beech can be used, but pine should be reserved for projects that incorporate its distinctive graining. Boxwood is best suited for endgrain carving. Other choices are pear, apple, plum, white beech, lilac, and hawthorn, as well as maple, soft birch, large-pored cherry, and elderberry.

Preparing the Woodblock. Woodblocks prepared for printing are not readily available, but a carpenter can do the job. The wood has to be thoroughly dry, well seasoned, planed by machine and by hand with successively finer blades, and finally smoothed with sandpaper. It is easier to handle soft wood after a hot solution of glue and water has been brushed on.

With a rag, apply a thin layer of fixative, shellac, or varnish to avoid swelling of the wood grain when you draw on the surface with paint or ink.

Endgrain blocks are more difficult to prepare. They often consist of several pieces that may have to be assembled on a lathe or with the help of a router. Smooth the block with a scraper and fine sandpaper, possibly with the addition of a little oil, and wipe with a shellac solution. Problem spots in the wood can be drilled out by a carpenter and replaced with a matching plug.

Transferring the Drawing. Draw directly on the wood (remember that your design must be a mirror image of the desired end result) or trace your design onto the block using one of the following methods.

1. Draw the original design on tracing paper. Apply a thin layer of varnish to the wood surface and dust with talcum powder. Put the paper on top face down, tape it in place, and rub. The drawing will appear in mirror image on the block.

2. Draw your design in mirror image on tracing paper and cover the back of it with light-colored chalk. Roll a layer of block-printing ink on the woodblock, dust with talcum powder, and prick the lines of the drawing through with a needle.

3. Draw your design in mirror image on tracing paper. Dampen it slightly and adhere it to the block with thinned waterproof glue. Rub the paper on securely, so that it will not separate from the wood during the cutting process. Use this method only for woodcuts; it is not suitable for endgrain carvings.

4. To transfer an image photographically, prepare the block as follows. To protect it from humidity, coat it with a layer of shellac. On a sheet of glass mix zinc oxide with the white of a fresh egg and rub the mixture onto the block with your hand. Let it dry and then sand it lightly with fine sandpaper. Next, apply a thin coat of photographic emulsion under low light. Let it dry in a dark place. Then put the film negative containing the design on the block, cover it with a sheet of glass, and secure it with clamps. Expose it to light; the time needed is a matter of practice and experience. To make the image on the block permanent, develop it with photographic fixer (hypo). Then rinse with water and let the block dry.

If your design is very detailed, protect the image on the block with a piece of paper that you can tape around the edges, and rip off the covering over each sections as you are working.

Tools and Techniques for Cutting. A woodcutting knife, such as the one shown in

Figure 425 is the best choice for carving woodcuts. It should be kept sharp and well-pointed, but only one side of the blade should be sharpened. Use a rough stone first, then a fine honing stone. Pull the knife when you cut. Depending on the length and depth of the cut and the hardness of the wood, hold the knife in your fist or like a pencil. Make your cuts slanted, never vertical or undercut (Figure 426). Turn the block to cut in different directions. Do not use the vee-shaped tool (Figure 427) to cut outlines. It is the right tool for correcting small mistakes or to perfect a corner.

Use a curved gouge (Figure 427) for small light areas that are cut against the grain. For larger areas use a flat chisel and a wooden mallet.

To cut letters in endgrain you will need gravers of various types, as illustrated in Figures 428 and 429. Keep them sharp enough to cut with only light pressure. Hold the graver with four fingers and rest your thumb, which guides the graver, on the woodblock (Figure 431). It is very important to rest the graver on a wooden wedge (Figure 432); this will prevent nicks and bruises on the surface of the block. Move the graver over the wedge in a lever-like motion. Should your tool slip and make an unintended cut, you can remove the spot with a punch and replace the wood with a plug.

If you are a beginner, try the different movements and cuts on a practice block.

Printing. Use only paper that is wood-free and soft and contains little size, and ink will adhere to it well. Hard papers or papers with a high size content do not take ink easily, and the ink may bleed. The best choices are printing or offset papers and handmade papers. For large prints with few details, use blotting paper; for woodcuts, use tissue paper.

Several hours before you start to print, dampen the paper by placing it between damp waste sheets, to make it more supple.

Use ink specifically made for printing. Use it full strength for paper with high size content and smoother surfaces, and dilute the ink with turpentine for more absorbent papers. Put the ink on a sheet of glass or a lithography stone with a spatula, and then use a gelatin roller to spread it evenly and transfer it to the woodblock. Work with small amounts of ink at a time and roll several times if necessary. Use a thin layer of ink for detailed artwork and a somewhat thicker layer for large-scale work.

Hand presses yield more precise and even results, and they also require less strength than prints that are made manually. Presses use less ink, yet the prints from them appear more saturated. Different presses are available for woodcuts and endgrain carvings.

Your paper should be larger than the printing surface of the block. Hold it by two opposite corners, with the rougher side facing the block. When the print is complete, lift the paper off carefully, starting from one corner. If you move too quickly, the paper may rip. At first the block itself usually absorbs a lot of ink, which may leave some empty areas on the paper: this should not tempt you to apply a heavier layer of ink for the next print. Instead, if the problem persists in subsequent printings, apply additional pressure to the problem spots or increase the pressure of the press.

If you have no access to a printing press, you can print by hand, using a bone folder, spoon handle, or other tool to burnish the paper. The best choice is a piece of pear wood shaped by a carpenter into a somewhat bent form, with rounded edges and corners, 12 inches (30 centimeters) long, 1 inch (3 centi-

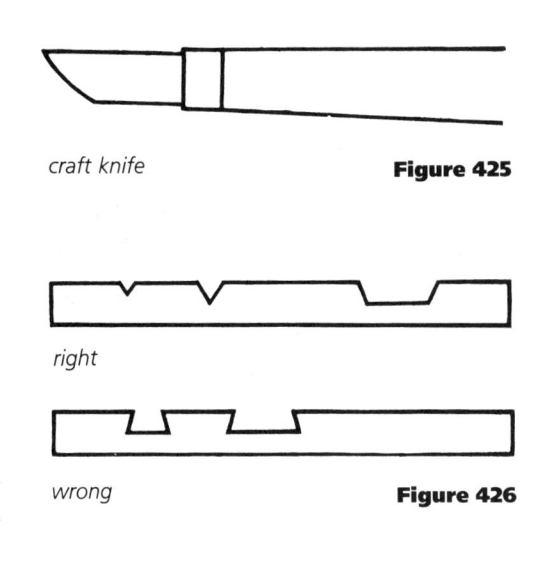

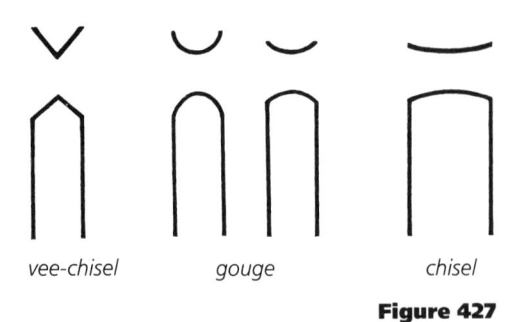

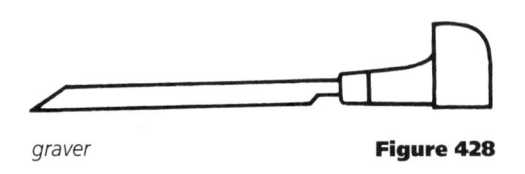

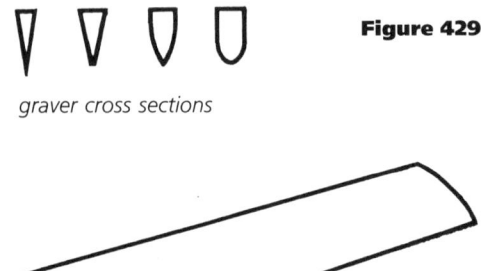

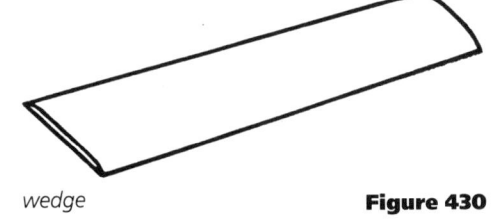

craft knife **Figure 425**

right

wrong **Figure 426**

vee-chisel   *gouge*   *chisel*

Figure 427

graver **Figure 428**

Figure 429

graver cross sections

wedge **Figure 430**

Figure 431

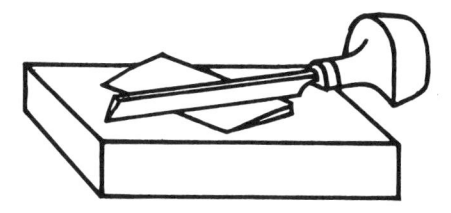

Figure 432

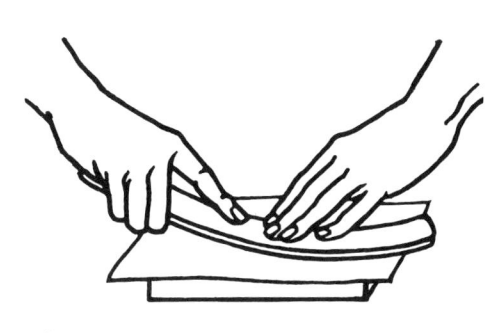

Figure 433

434 *Weight, right angle, and paper.*

meters) wide, and ¼ inch (7 millimeters) thick. All its surfaces should be smooth and polished with wax or varnish so that it will glide easily across the paper.

To put the paper on the block, hold it with both hands and set the left side of it down first. Hold it in place securely with your left hand while you gently stretch the rest of the sheet on with your right hand. Keeping your left hand in place, with your right hand smooth the paper from left to right until the entire surface is in contact with the block, then press it on lightly with your palm. Before you use the wooden burnishing tool, protect the paper with a sheet of wax paper or oiled paper. Move the tool in small circular motions from left to right and use the edge of it if more pressure is desired. Position your left hand so that the paper cannot be crumpled.

Clean the block with turpentine after each printing: it will be destroyed if you allow the ink to dry on the surface. Corrections are often necessary after the first printing. If the artwork on the block is difficult to see after you clean the block, dust the block with talcum powder and wipe the excess off with your hand.

If you want to make simultaneous prints of woodcuts and other typographic materials, be sure to use blocks of the same thickness.

Color Printing. Each color requires a separate block. To avoid misregistration during the transfer of the design to the the blocks, print the first block onto all others and cut away everything that appears black as well as everything that will not appear as second or third color. You can use needle registration or an angle-guide setup.

For the first method, make two needle-point holes on the surface of the first block. Make sure that they are visible on the print and punch holes at these spots in the paper with the needle. Transfer this print of the first block onto the next block, make the new print, and from the back side of the paper punch through the holes with your needle into the wood. In this way all blocks will be marked with two dots at exactly the same points, which will serve as registration guides for the paper.

To use an angle guide (Figure 434), you will need two pieces of wood at a right angle. Now make a print of the first block and do not take it off. Turn the block upside down—the paper will stick to the surface—and place it on a level piece of cardboard. Place your guide at one corner of the block and mark all positions. Hold the guide, remove the block, and put the next block face down in its place (the paper should remain on the cardboard) and make the second print.

A third method is to make one print of the first block for every additional block on tracing paper and glue the sheets onto the blocks.

Linoleum Cuts

Linoleum cuts are prepared in much the same way as woodcuts. Linoleum and similar floor coverings can be used as long as they are at least ⅛ inch (3 millimeters) thick; otherwise it is not possible to cut out large areas. Choose materials that contain no sand, will not crumble or crack, and have no streaks or bubbles. Reinforce the linoleum with a cardboard backing and back the cardboard with another layer, so that the adhesive does not warp the backing.

Transfer your design to the linoleum and spray it with fixative so that it does not get rubbed off. If the traced lines of the design are imprecise and your design calls for extremely precise letters, use a sharp-pointed pencil to make corrections, but do not press too hard or you will mar the surface of the linoleum.

To cut outlines, do not use the vee-chisel but rather a knife. This is more work, but the contours will be more precise. To cut the interior spaces, use a U-shaped gouge. There are many tools designed for cutting linoleum: use one with a short handle that allows a secure grip. Do not use tools that are designed for woodcuts. When you work with a knife, pull towards yourself; gouges should be pushed away. Make shallow cuts of about 1/32 inch (1 millimeter) depth.

Before you print, you can temper the linoleum block with alcohol or methylated spirits. Use ink made for linoleum printing. If you do not have access to a press, make your prints with a burnishing tool such as those described for woodcut printing. After printing, clean the block with alcohol.

Engravings

Engraving is a good technique for small projects that require precision, such as initials, monograms, and ex libris plates. Here the image to be printed is cut *into* the surface of the plate, rather than raised above it. Both material and technique demand exactness and well-defined forms. Plates can be of various metals — copper, zinc, brass, and others. The plate should be about 3/16 inch (4 millimeters) thick if it is to be used in a printing press or with other type.

Tools are gravers, burins, needles, and scrapers. A magnifying glass, if necessary, and a honing stone should be at hand.

To transfer the design, clean the plate and cover the plate with a thin layer of gum arabic; then copy the drawing or trace it on with carbon paper. Define the outline with an engraving needle and remove the burr with a scraper. Repeat the process for deeper lines, or cut them with a graver. Afterwards widen the lines as desired with a graver, which should be guided in a rocking motion over a wedge, as in woodcutting. Rest the plate on an engraving block or pad and turn the plate to the optimal positon for each cut (Figure 435). Keep the graver sharp during the process: use the honing stone frequently. Use a scraper to remove all burrs before printing. All remaining problem areas will be visible clearly after the first print. Scratches can be removed with a burnisher or scraper. Any resulting indentation in the plate can be corrected by hammering the plate from the back; the plate must be placed on a completely flat metal surface, with a thin piece of cardboard between the two to protect the plate. If you have more substantial errors to correct, solder additional metal to the surface and recut.

Before printing, trim the plate precisely and either nail it to wooden crosspieces or glue it to metal ones.

Resist Technique

The resist technique is a good way to make negative calligraphic forms. You will need gum arabic or another water-soluble adhesive. Thin a small amount of it with water in a bowl and add watercolor to make the design visible. Check the consistency of the mixture before you start: if it is too thin, the resist will not work; if it is too thick, it will clog the nib of your pen too quickly. Even under ideal conditions you have to clean your nib frequently. Apply the resist material to the nib with a brush, as if you were dealing with ink. When dry, the letters should be shiny but not too thick, or they will crack.

Next, using a rubber roller, cover the entire paper surface with the desired color of water-insoluble paint or ink. Repeat the process until there is an even

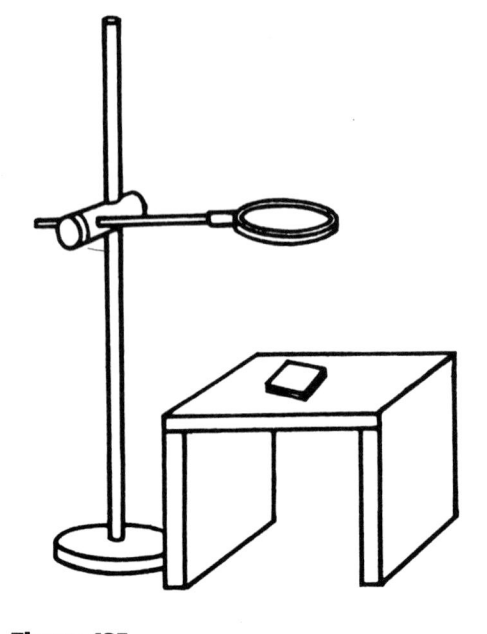

Figure 435

layer of color over the sheet. Then hold the paper under running water to dissolve the adhesive, thereby removing color in the resist area. You may have to wipe it gently with an old brush. Stretch the damp sheet on a board to dry.

The letters created by resist often retain traces of color in fine crack lines, and may have slightly uneven contours: both effects can be quite attractive.

An alternative method for applying color to resists is to use paint thinned with turpentine and a bristle brush or a stippling brush. The outline of the letters will be fuzzy, because the thinned paint will eat away at the resist, and more color will remain in the washed out areas.

Scratch Technique

Scratching letters on glass or sturdy acrylic is a very useful technique for designers. Add some glycerin or soap to your paint to keep if from cracking; too much will cause the color to streak when you scratch it. Place the design under the surface. Draw the pattern with a brush, and scratch its outline with a sharpened lithography scraper. It is easy to correct the drawing.

The scratched design is more clearly defined than a pencil drawing on paper, and can be reproduced photographically.

You can prepare a simple scratchboard with smooth cardboard by applying a white base made of carpenter's glue, whiting, and some zinc oxide. Apply several layers and smooth with very fine sandpaper. Then apply a layer of fixative to keep the ground separate from the ink, and cover this with black Chinese ink or printing ink. Use a lithographic scraper or a graver to scratch in the design. Cover errors with fresh ink, and rescrape. Again, the scratched artwork can be reproduced.

Lettering in Architecture and the Environment

General Remarks

There are well-known connections between styles of architecture and lettering. Until the dawn of neoclassicism, all inscriptions were matched carefully and harmoniously to the building they were intended for. During the nineteenth century a general decline in taste and the rise of advertising caused many changes. On today's streets we find but few examples of lettering that is used beautifully and in harmony with the surrounding architectural elements. The prevailing reliance on technology and formalistic amateurism proves that neither architects nor the companies that execute the signs are able to handle the problems of lettering.

Better models could improve the situation substantially, but even those will not suffice for important design projects, where only well-educated graphic artists can do what is necessary. It is important to remember that lettering of any kind in connection with architecture must never appear as an afterthought, but has to be a vital ingredient of the total design. Lettering should not only satisfy the immediate demands of the particular purpose: it should also satisfy aesthetic and artistic demands for years to come.

It is equally important to consider the optical effects of the lettering arising from the building's position or traffic situation. The placement of the lettering and its size depends on all these factors as well as on the length of the text and the letter style.

When you plan a sign for a store, consider what will be in the shop windows. If you are dealing with an building of historic interest, incorporate the sign in the window area and avoid interfering with the facade. Lettering should not overpower architecture—it is usually much too large.

In choosing materials and techniques, take into consideration the lighting conditions both day and night, the way in which the lettering will be secured on the structure, requirements of durability, and last but not least the colors and textures of the background and the surrounding areas. When choosing a letter style, remember that on modern structures we often find a constructed sans serif. The effect of this style of lettering parallels a functional puritanism in modern architectural design. Basic attitudes in architecture have been changing, but have not yet influenced lettering and its application. But those style problems aside, roman types such as Garamond, Bodoni, or Walbaum and historical sans serifs or Egyptians are often much better suited to modern building styles because they enliven the glass, aluminum, and concrete surfaces in a most appealing way.

It is more difficult to choose the right lettering style for a historic building. Proceed with great care. The letters must be compatible with the era. Neon signs have no place on architecture that predates electricity. The eclectic styles of the latter half of the nineteenth century pose special problems. Here, keep neon signs small to avoid stylistic contradictions, and consider the variations of classic typefaces that were developed at the beginning of the nineteenth century and the beautiful and detailed rich Egyptians and sans serifs from the same period.

Whatever the style you select, it must lend itself to the method and materials that you want to use. Not every lettering style can be adapted. Handwritten models are particularly sensitive. Neon signs could represent handwriting, as long as the letters are mounted on a flat surface and not on three-dimensional forms. Three-dimensional lettering should not be used if the viewers' vantage point will be low, since the perspective may distort the forms. Problems can arise if an already existing logo has to be incorporated into the design; that is why it is best to plan such applications when you design a logo.

A good relationship with the architect and the workmen is essential for success in a construction job. It is imperative to obtain all relevant technical information before the designing process can begin.

Make a sketch of all the characters in the right proportions. Get a photograph of the building you are working on, and make a model in cardboard, if necessary. Use the actual colors, if you can, and check them on location. It might even be possible to project a slide of your design onto the appropriate area of the building at night to get an impression of the final effect.

Practical Hints

Letters can be scratched in plastered walls (sgraffito), painted, set in mosaic, or constructed in various materials and applied. Match backgrounds of ceramic or natural or artificial stone with letters of the same material. The lettering can consist of single letters that are flat or three-dimensional, or of completed signs that are attached to the wall.

Material for single letters can be metal that is treated in various ways (plated, gilded, chromed, enameled). Letters with serifs are usually cast or punched. If every letter is produced individually, it can be cut from sheet metal or from metal rods, formed, mounted, and assembled. Welding is another possibility. Plastic materials can be poured into forms, reinforced with glass fibers for

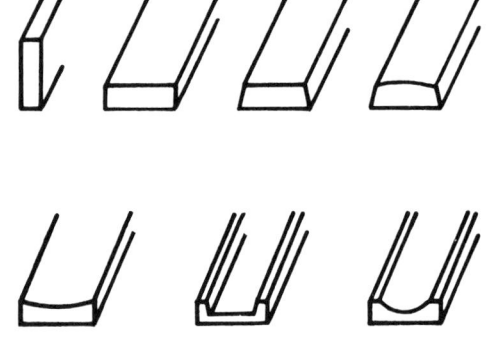

Figure 436

Figure 437

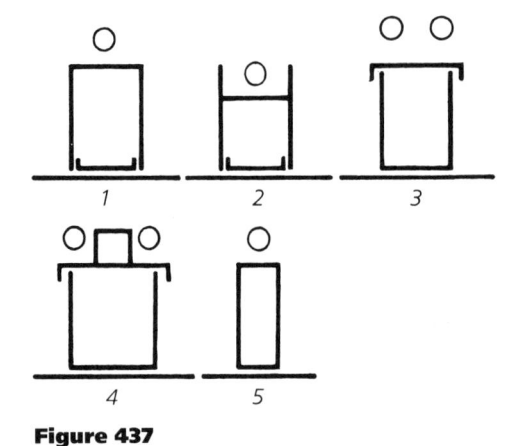

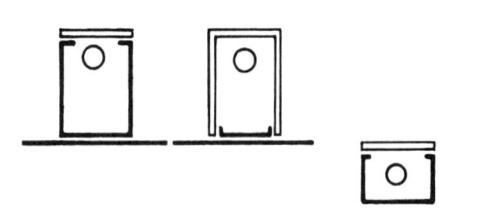

Figure 438

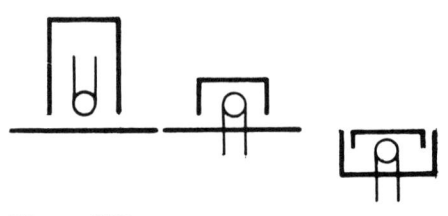

Figure 439

larger formats, or cut. Wood letters can be cut out with a saw and painted.

Hollow shapes can be created from glass, plastic, or sheet metal. They can be sawn, bent, and mounted.

If you attach single letters to a wall, mount them at a slight distance from the wall so that rain will not create dirty streaks on the wall. Figure 436 shows some profiles of relief letters.[13]

The backgrounds for signs can be made of sheet metal (bronze, steel, and other metals), glass, plastic, wood, and stone. On metal you can paint letters, punch or saw them out, or use a relief

---
13. From Walter Schenk, *Die Schriften des Malers*. Giessen: Fachbuchverlag Dr. Pfannenberg, 1958.

440 *Daytime.* **441** *Nighttime.*

442 *This is an interesting idea, but unfortunately the details of the lowercase letterforms cause light spots. Capital letters might have made a better solution.*

technique, such as bronze casting, etching, welding, inlay, or even gluing. On glass you can paint, engrave, or etch.

Letters chiselled into stone are appropriate for historical buildings and commemorative plaques.

Concrete is a good medium for relief or three-dimensional work—it is not used nearly often enough in this capacity.

Always make certain that the letter type you choose can be adapted for use on your materials. Capitals are preferable to lowercase because they are more static and monumental in their appearance.

The list of possibilities is by no means complete, nor are the following descriptions of specific techniques. Only the aspects that influence the design decisions or that can be executed by the graphic artist alone are discussed. Whatever your choice, consult with specialists before the work commences.

Special Techniques

Neon Signs

Neon lettering poses special problems because it has to remain visible day and night, and the character of the letters should not change under different light conditions, especially if a logo is involved. If the tubes are attached to raised letters, their surface should not be too striking. Consider the color and the intensity of the light as well as all design elements in the immediate surroundings. You can mount the tubes on three-dimensional letters or directly on a colored background, but in the latter case your letter spacing must be ample enough to avoid blurring. Figure 437 shows possible arrangements of tubes on three-dimensional letters.[14] Variation 2

---
14. From Gottfried Prolss, *Schriften für Architekten* (Lettering for architects). Stuttgart: Verlag Karl Kramer, 1957.

---

is especially useful, because the protruding edges keep blurring to a minimum. In variations 4 and 5 the surface of the raised letter is indirectly illuminated, so the letters appear as silhouettes. None of these five versions looks its best during daylight hours, because mounting the tubes on raised letters creates a certain restlessness. Aesthetically much more pleasing are constructions where the tubes are placed inside three-dimensional letters and covered by a translucent sheet of glass or acrylic. Good effects can be created with side walls of contrasting color. Since acrylic can be shaped easily, it is possible to make the entire structure except the base plate out of it. Three-dimensional letters with diffuse light effects can be mounted on a wall or recessed into it totally or partially. Figure 438 shows the most common arrangements.

A third variation is to light the letters indirectly. Here the three-dimensional letters are open at the back and illuminated by tubes that are recessed into the wall. The letters appear as silhouettes (Figure 439).

A special form of either direct or indirect lighting consists of rectangular, oval, or round shapes which totally enclose each transparent letter (Figures 440 and 441). Signs that are mounted on roofs have to be visible against bright or dark skies; the mounting hardware should be as unobtrusive as possible.

Technical requirements for neon letters can be quite complex, and it is impossible to comment on all of them in detail. Problems with materials can stall a project completely, so it is wise to gather all relevant information from appropriate companies before any designs are chosen.

Lettering on Walls

The range of choices is determined by the type of material at hand. Lime, cement, or mixed grounds, if they are new, can be treated in fresco or sgraffito technique. The structure of the surface should be as smooth as possible. In industrialized areas air pollution will attack the surface of frescoes and eventually destroy them. Match colors to the content of the ground—chalk, latex, oil, emulsions, and mineral pigments all have specific requirements, and their proper management falls in the realm of painters. For high-quality lettering it is necessary to collaborate with painters, who are not usually trained in lettering design, and combine the specific skills of graphic designer and painters. The graphic designer will be responsible for the choice of type and the arrangement of letters and will produce a design as well as a pattern drawing on paper, ideally in full size. He or she will also discuss the colors and their technical requirements with the workman, and assist in the mixing of color samples.

Sgraffito.[15] Of all mural techniques, sgraffito is described in detail because it requires more than a simple transfer of the design onto the ground, and it is difficult to find specific instructions for the process.

The earliest sgraffiti (*sgraffiare*—to scratch) date from the Renaissance and are found in northern Italy as weather-resistant murals. The German architect Gottfried Semper reintroduced this nearly forgotten technique in the nineteenth century for decorating the outside of buildings. Modern architecture avails itself of this technique for lettering on facades or interior walls. It is, however, necessary that the entire wall surface be applied at the time of lettering.

The technical requirements must be considered from the start. No intricate details are possible; fat-face types cannot be used because their thin strokes do not withstand the weather well. Consider the optical effect of shortening in perspective, which can change the appearance of letter parts, especially when they are thin.

It is best if the graphic artist personally transfers the design onto the wall rather than leaving this job to workmen. Beginners are advised to practice cutting and scraping on sections of plaster. The preparation of the wall itself and the necessary material are best left to the trained craftsman. The ingredients of the plaster are slaked white lime and sand, preferably washed river sand. The sand should not contain clay and other impurities, which may affect the firmness and color of the plaster. Lime and sand are mixed in the proportions 1:3, with slightly more lime for the first layer than for succeeding ones. A small amount of cement is added for exterior work. All pigments used have to be lime-proof: test before you start work by mixing the paint with white lime and exposing it to sunlight while wet. Some pigments may contain ingredients that cause the plaster to dull or bleed or develop a whitish film. Unadulterated pigments are the best choice. Many oxides are on the market, but experience shows that earth pigments last longest. These circumstances restrict the available palette to the following colors:

Yellow: yellow ocher, sienna
Red: burnt ocher, burnt sienna, English red, red bole
Brown: umber
Blue: blue verditer
Green: green earth, green umber
Gray: slate gray
Black: vine black

Do not use more than one part pigment to two parts plaster, for exterior or interior walls.

The tools used are wire loops and implements for scratching in different widths, and these are available in stores or can be improvised or made by any smith (Figure 443).

Apply only as much plaster as you can finish working in one session, and protect outside walls during this time from sun and rain with a tarpaulin. If you have a choice, do your work during spring or fall when air humidity is high, and avoid work during very dry periods or under intense sun.

The actual steps are as follows. A base layer of mortar is applied to the wall, with areas that are designated for the lettering left half the thickness of the other parts. The base layer must be left to dry thoroughly and rewetted before the colored layer, about ¼ inch (5 millimeters) thick, is applied. If additional layers of color are desired, about two hours' drying time should be allowed for each layer. Finally, the layer of facade plaster is applied. During the drying process a thin transparent skin, or sinter, will form on the plaster and the color will brighten, which makes it possible to create letters by scratching off the lighter top layer even if no colored layer was placed underneath. This sintered layer also makes later corrections impossible. If you transfer your design in the traditional way with charcoal and pounce bag, the resulting dots will be incorporated in the sinter and have to be removed during cutting. It is therefore better only to press the lines through. If the work time exceeds ten hours the surface may begin to chip. The sgraffito is done either by cutting the letters into the plaster, to make recessed colored letters, or cutting away the surrounding area, creating

---

15. Dr. Gerhard Winkler, Leipzig, was consultant for this section.

white raised letters on a colored background. The scratching should only be deep enough to expose the top of the colored layer. The lower edges on letters that are exposed to the elements must be beveled to keep water from collecting in the corners (Figure 444).

Lettering Design for Stone Carving[16]

Lettering on stone for memorial tablets, gravestone markers, and plaques on public buildings is the specialty of stone masons. Unfortunately most of them have little or no education in the field of type design and often choose the worst of a limited and outdated selection of patterns. The graphic designer, on the other hand, often has no concept of the special requirements of the material or of the chiselling process. Many drawn or written letters cannot be transferred to the medium of stone, and what looks good on polished granite might be impossible on travertine. The following instructions cannot turn a stone mason into a graphic designer or vice versa, but they might further understanding of basic procedure to make productive cooperation possible.

Best suited for letters cut in stone are the letters that we find on monumental inscriptions—roman capitals, sans serifs, and faces with little contrast in the stroke widths (Figure 445a). The capitals of Gill Sans are a good example, if the M is slightly modified. Frequent models are the capitals of the Trajan Column with their pointed serifs and light contrasts in stroke width (Figure 445b). Other good choices are capitals of other condensed sans serifs if the stroke width is fairly narrow. Both raised and recessed versions are possible. The use of roman

---

16. This section was written in cooperation with the sculptor Fritz Przibilla, Leipzig.

---

lowercase, however, is questionable, because these letters were developed as bookhand, and the chisel would have to imitate the movements of the pen. Lowercase roman letters are neither simple enough to be monumental, nor rich enough in detail to be decorative. All other styles that developed their characteristics from the motion of the pen stroke are equally unsuited to transfer into stone.

A possible exception are the baroque italics and fraktur styles of the thirteenth to seventeenth centuries. They can be found on old grave markers and have great decorative appeal. Their details lend themselves to the expressive range of the stone mason's tools. The stones themselves were often round in shape and their surfaces slightly convex. Textura was usually worked in raised letters, fraktur and italic recessed.

Old graveyards and churches are rich sources of inspiration for the student of historic lettering styles. Rubbings can be taken by taping paper over the stone and rubbing with the flat side of a piece of chalk or wax crayon over the letters until their shapes become visible on the paper.

The following alphabets in this book are suitable for work in stone: Figures 82, 102, 188, 189, 351, and 358.

When you design a text for a stone cutting, keep in mind that your tool is a chisel with a specific range of possibilities. Draw the contours of the letters only on tracing paper and leave detail forms to the mason. Specify the format and the placement of the text on the stone. Short inscriptions look best in a symmetrical arrangement: longer text should be set flush left. Let the mason transfer the text onto the stone.

There are two categories of stone: soft stones such as sandstone, limestone, and marble, and artificial stone materi-

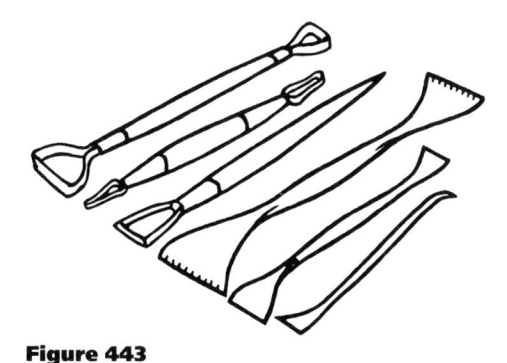

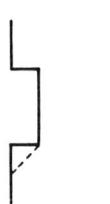

Figure 443

Figure 444

als, and hard stones such as the various forms of granite.

The use of highly polished stones for grave markers is controversial among experts, and many prefer the matte surfaces to the polished ones.

Figures 446 and 447 show different profiles of cut letters.

Soft stone. Figure 446a has a polished surface and large areas are scratched, b has a polished or scratched surface, c and d have a polished surface and a tooled ground.

Hard stone. Figure 447a has a polished surface. Use larger letters without serifs on scratched surfaces. b can have a polished or scratched surface and a tooled ground. The profile in c is seen frequently and results in letters that look as if they were superimposed on the stone. The connection between ground and letters must remain evident.

It would be of little use to explain the various techniques involved in any detail. The interested graphic artist can observe a stone mason in his shop and learn about the appropriate use of his tools.

The depth of the relief depends on the size of the letters. Raised styles are often too high and appear clumsy. If a letter is 1 to 2½ inches (3 to 7 centimeters) high, a relief height of ¼ inch (3 to 5 millimeters) is enough. Consider porosity, color, pattern, and structure of the stone surface and practice single letters on small scraps of stone. Estimate the effect of light at the future location of the stone and the relatively quicker weathering of softer stone.

Recessed letters are often gilded or otherwise colored for better legibility. This is a practice that was popular among the early Greeks and during the Gothic and baroque eras. Do not use silver or other metals of similar color—they look alien, even ugly, in combination with stone. Other colors should be chosen slightly darker than the stone itself and in the same shade. Use earth pigments for soft stone and artists' oil paints with some added oil for hard stones. The best solution is lettering designed to be legible without the addition of color, and the worst effects are without a doubt created by two-color enhancements that imitate light and shadow effects on the two sides of a V-shaped profile.

Neither paint nor gold will permanently adhere to marble that is exposed to the elements. Indoors the light situation might be controllable to make additional color unnecessary. Marble can be used for finely detailed lettering styles. Gravemarkers with bronze plaques or letters can be seen frequently, but the bronze oxidizes after only a few years and will in time destroy the surrounding stone.

Raised or recessed lead letter, on the other hand, can be used on polished or scratched surfaces. Use a relief height of ⅛ inch (2.5 to 4 millimeters) for 4-inch (7- to 10-centimeter) high letters. The profile can remain smooth or be decorated with grooves. Weathering will not occur (Figure 448a). Another technique that can be used if the lettering is not too ornate or detailed is inlay, which requires great skill and is very time consuming. Use light marble or other light stones. Chisel the letters in a profile that is slightly wider at the bottom and drill three or four slanted holes for each letter (Figure 448b). Depending on the properties of the stone the lead strips of the chosen profile are either hammered in or melted and poured in, and later polished until stone and letters are perfectly level.

Letters cut in wood require similar aesthetic considerations, but are executed with different tools in a different material.

Figure 445a   **Figure 445b**

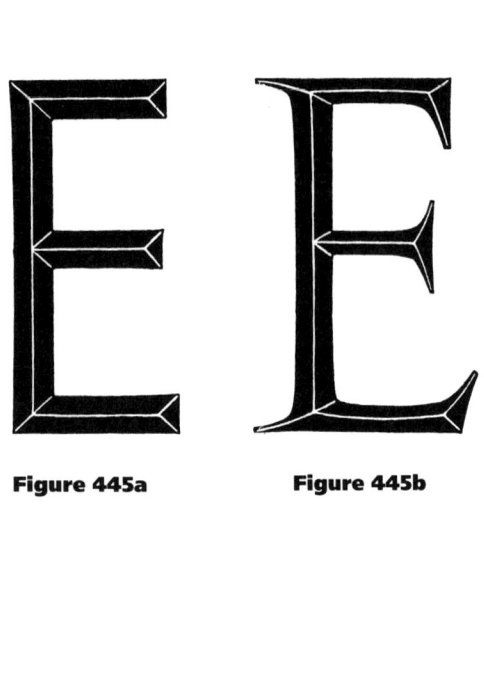

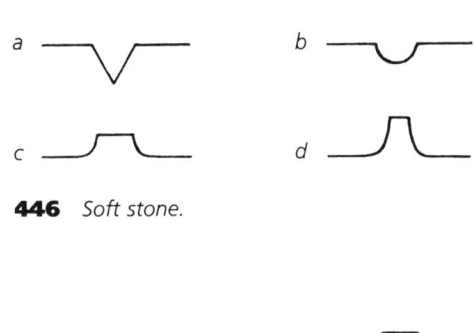

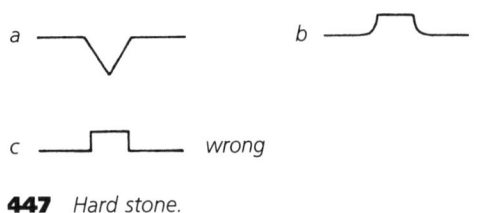

a   *b*
c   *d*

446 *Soft stone.*

a   *b*
c wrong

447 *Hard stone.*

Figure 448a

Figure 448b

drilled holes

Figure 449

450 *Main street in Leipzig during the Spring Fair, 1969. (Photo by Doris Beinecke.)*

Lettering in Bronze

Raised letters look better than recessed ones on cast bronze plaques. Transfer a mirror image of your design onto a plaster base and cut the letters out with a knife or other woodworking tools (Figure 449). For small plaques, consider using wood, soapstone, or slate. The relief must be all the same height. Alternatively, you can cut out plywood letters and fix them to a base as a pattern. A foundry will use your model to make a sand mold for casting.

Lettering on Glass

Lettering on glass is frequently used in advertising and can be found in windows, house fronts, and exhibitions. The process actually belongs in the domain of the painter or signpainter, but since painters are not usually trained to work with type, it is best to refer involved projects to a graphic artist to achieve aesthetically pleasing results. In the design of exhibition spaces it may be most effective for the graphic artist to carry out the entire project. If more than one sign is to be produced, silk screening is usually the technique of choice.

For this technique, the letters that are to be printed are cut out, and this can only be done if the form of the letters is appropriate. Quickly sketched brushwork, for instance, is not suitable. Use only lightfast color, and avoid large light areas next to large dark areas in designs that will be exposed to intense sunlight. Colors react differently to exposure, and expansion could cause cracks. If the sign is transparent, the contours of the letters will be somewhat blurred: to minimize this effect, keep plenty of space between letters and draw a pattern in full size on tracing paper to check the effect.

A short description of the process follows for the designer who wants to execute the work completely. Metal foil used to be employed, but has now been replaced by paintable foil. Clean the glass with alcohol and make sure that there are no scratches on it. Then apply the foil and let it dry in horizontal position. If the foil is translucent put the drawing on the transparent paper in mirror image on to white drawing paper and place the glass plate on top. Now cut around the letter with a single-edged woodcutting knife or a similar cutting tool. If you want light letters on a dark ground, it is best to remove the foil that covers the background area and leave the letters standing. If you want dark letters on a light ground, cut around the letters and then remove the letter shapes.

Apply the color with a stippling brush. After the first color has dried, remove the rest of the foil. If you want background color, apply it with a rubber roller or stipple it on. The paint should not be too thick. Half oil paint with 35 to 40 percent French turpentine and true linseed oil varnish are best. Addition of several drops of thickened linseed oil is absolutely necessary, but if you add too much blisters will appear in the paint when it is exposed to the sun. Latex paints are durable, but require the application of several layers because they do not cover well.

Large-scale Lettering on Banners

Large banners and posters make striking advertisements for special occasions, but you must consider the architecture of the surrounding buildings and spaces. Do not cover or obstruct windows, and confine your signs to store windows if you are dealing with old buildings of architectural interest. Banners stretched across a street are a good solution. Figure 450 shows the effect of large vertical banners that are used to decorate Hainstrasse in Leipzig during the annual fair.

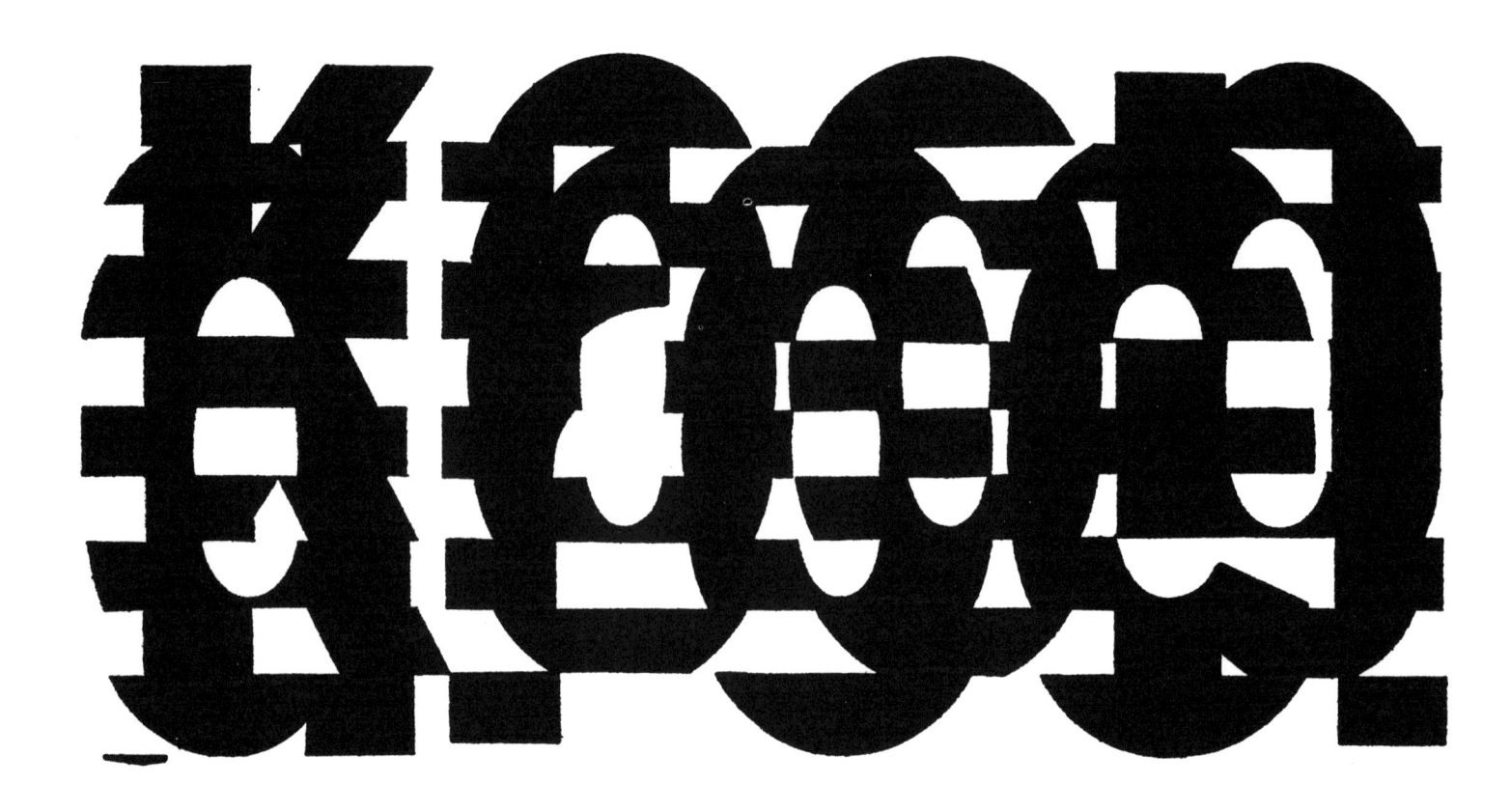

Large-size advertisements are usually attached to buildings or are placed in certain locations at streets and plazas. Permission of the municipal government or the police may be necessary. The proportions should always be matched to the architecture of the area.

Even a simple banner requires a full-size sketch that takes borders and spacing into account. All measurements for lines, letters, spacing, and letter widths are then transferred with charcoal onto the poster or banner.

Resist the temptation to create something "artistic," and use solid models instead. Best suited are to these applications are condensed or normal sans serifs, classical roman, Bodoni capitals, and other neoclassical faces. A wide italic can only be used if a wide enough marker is at hand. Do not use wide letters unless you have a wide enough marker to make them with a single stroke. If you have to draw very large letters, make templates of the most commonly appearing crossbars, curves, serifs, and other elements. Assemble the templates, trace them with a pencil, and draw the contour with the help of a mahlstick. Then fill in the letters with paint. If you want to use formal or very detailed styles, enlarge a smaller pattern or a photograph by overhead projection.

451 *Composition with type elements, by Franz Mon.*

CHAPTER
5

Portfolio of Type and Lettering in Practice

454 *Calligraphy by Raymond Gid, 1966. From an exhibition catalog of the Association Typographique Internationale, Paris, 1967.*

455 *Mark for the magazine Die Woche, by Otto Eckmann, 1897.*

456 *Calligraphic initial by Imre Reiner.*

452 *Calligraphic text from Paul Eluard's poem collection, I am not alone, by Irmgard Horlbeck-Kappler.*

453 *Calligraphic study by Albert Kapr.*

Figure 454

Figure 455 **Figure 456**

215

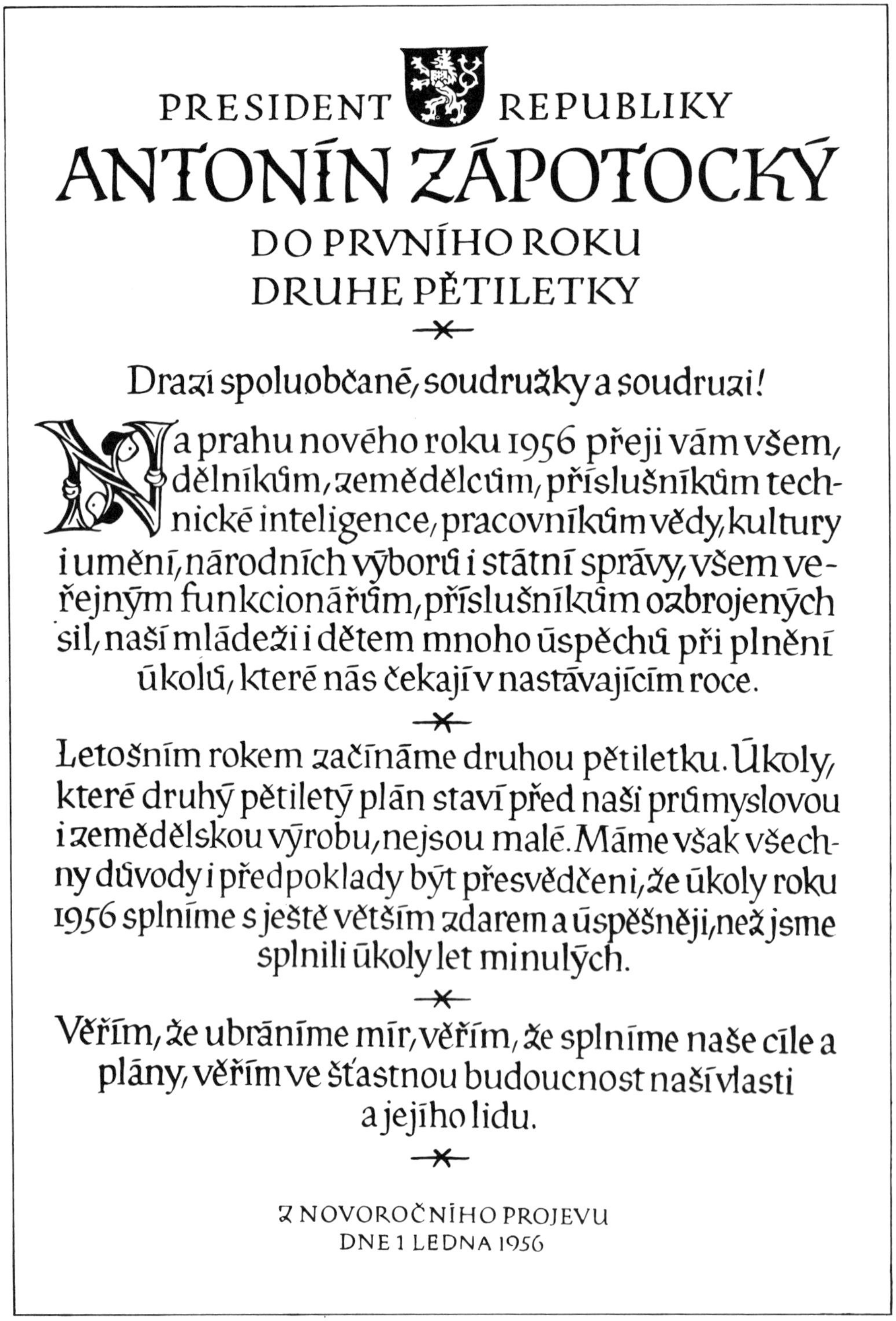

Figure 457

457 Graphic design for a sample page (reduced), by Oldrich Menhart.

458 Lettered name, by Georg Trump.

459 Lettered name, by the author.

460 Lettered name, by the author.

461 Lettering from an advertisement. From Jan Tschichold's Treasury of Alphabets and Lettering. Reprint. Copyright © 1952, 1965 by Otto Maier Verlag, Ravensburg.

462 Monograms by Oldrich Menhart.

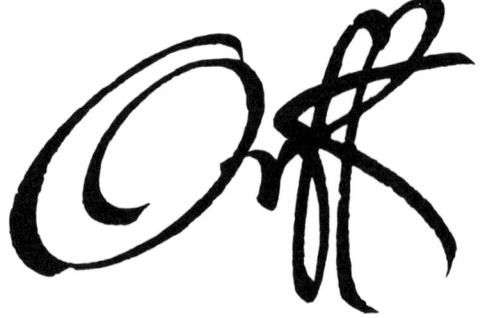

Das größeste ist das Alphabet,
denn alle Weisheit steckt darin.
Aber nur der erkennt den Sinn,
der's recht zusammenzusetzen versteht

Geibel

Figure 463

Figure 464

463 *Calligraphic text (reduced), by Hermann Zapf.*

464 *Title with typographic decoration by Imre Reiner. From* Lettering in Book Art. *St. Gall, 1948.*

217

UCASSIN WAR AUS BEAUCAIRE,
Einer Burg von reichem Leben,
Doch der schönen Nicolette
Nicht vermocht' er zu vergessen,
Wie der Vater ihn auch schalt.
Und die Mutter sagte dies:
„Junger Tor, was willst du wagen?
Lieblich zwar ist Nicolette,
Doch verkauft ward sie von Heiden,
Die sie aus Karthago raubten.
Willst du eine Gattin nehmen,
Nimm ein Mädchen hohen Standes!"
„Mutter, das vermag ich nicht.
Edler Art ist Nicolette,
Hold von Antlitz und Gestalt;
Ihre Schönheit wärmt mein Herz.
Nicht mit Unrecht muß ich lieben:
Keine ist so auserlesen."

465 *Page with woodcut initial by Robert Jung. From* Aucassin and Nicolette. *Vienna: Avelun Presse, 1919.*

466 *Title page by Imre Reiner. From* Lettering in Book Art. *St. Gall, 1948.*

467 *Title page. Lettering by Albert Kapr, design by Horst Schuster.*

468 *Title page for Thomas Mann's* Thamar, *designed by Gunter Böhmer. S. Fischer Verlag, n.d.*

469 *Calligraphic design for a chapter opening, by S.B. Telingater.*

470 *Page from a book on Karl Schmidt-Rottluff with woodcut initial capital by Will Grohmann, Stuttgart, 1956.*

471 *Lettered initials by Imre Reiner.*

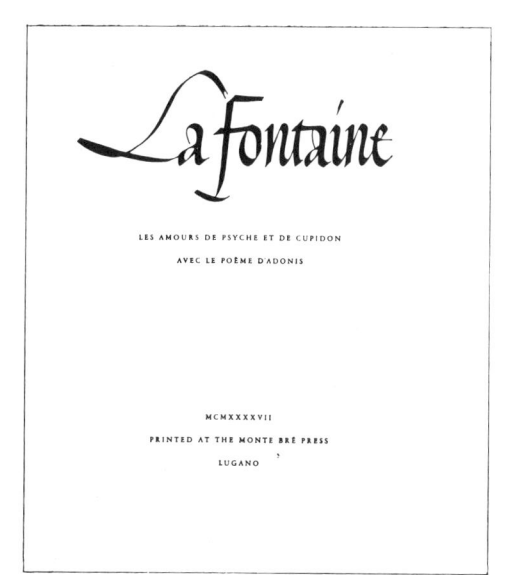

Figure 466

Figure 467

Figure 468

Figure 469

Figure 470

Figure 471

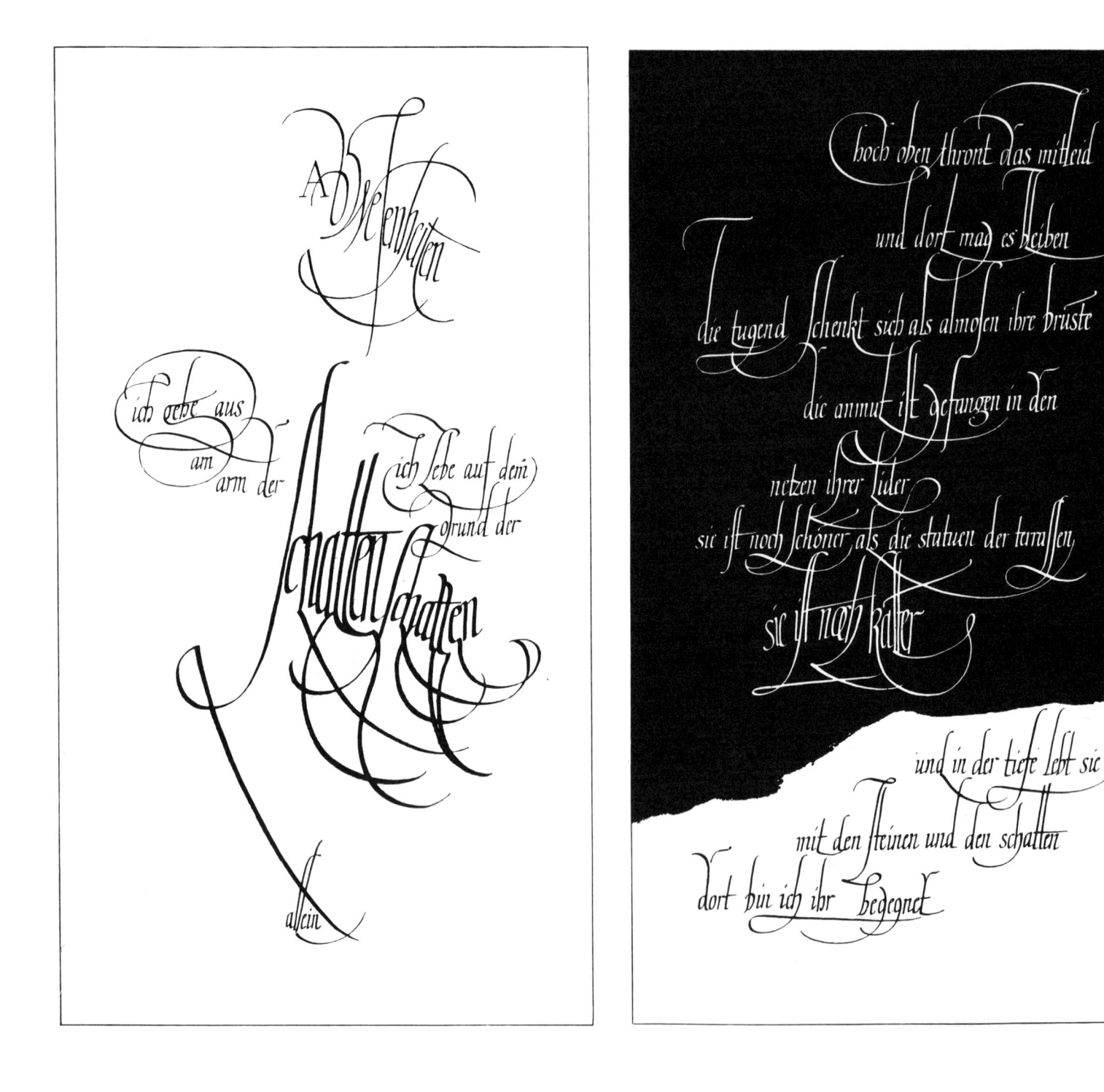

472 *Spread from Paul Eluard's* I am not alone, *by Irmgard Horlbeck-Kappler. Leipzig, 1965. Original size, 220 by 384 millimeters.*

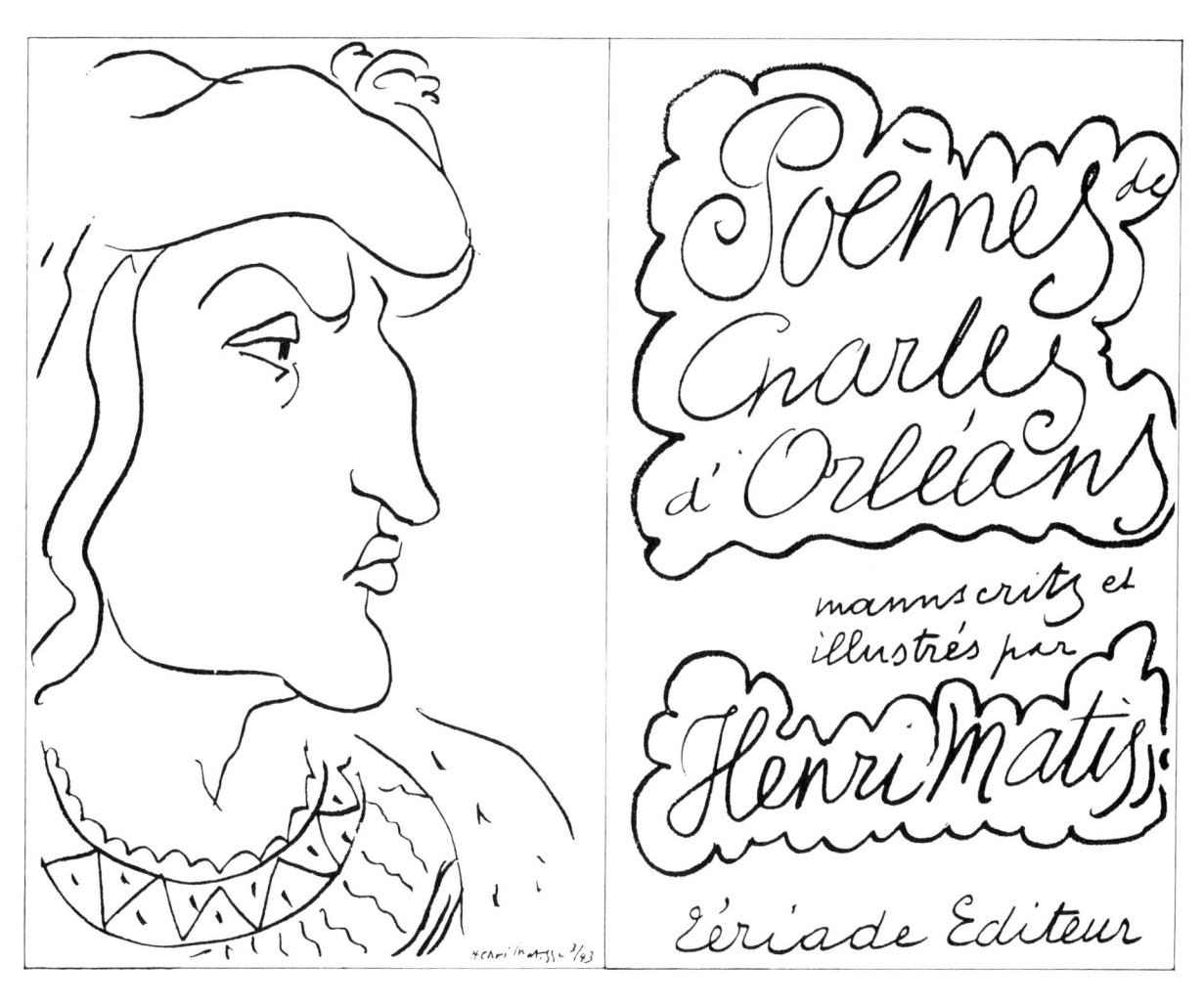

473 Lettered and illustrated title page to Poèmes de Charles d'Orléans by Henri Matisse. Paris, 1950. (Photographed from a multicolored lithograph. Original size, 265 by 410 millimeters.)

474 Handwritten and illustrated engraving from Gongora, by Pablo Picasso. Volume 1 in the series "Les grands peintres modernes et le livre." Paris, 1948. Original size, 243 by 350 millimeters.

Figure 475

Figure 476

Figure 477

Figure 478

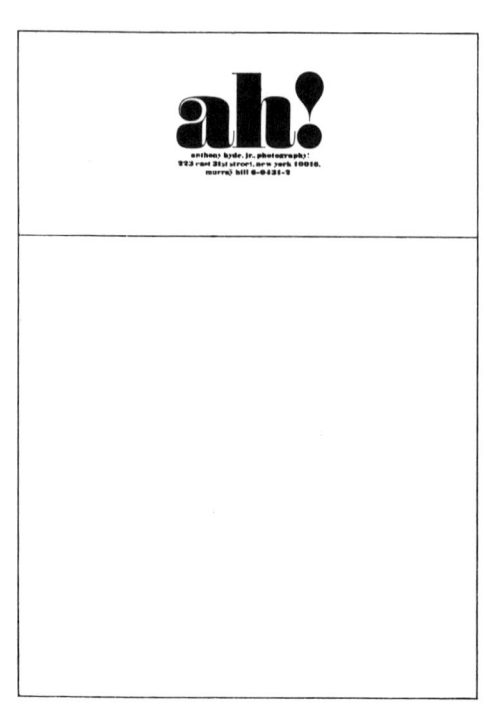

475 Letterhead with calligraphic logo. Design by Bertil Kumlien. From J.W. Zanders, Der Briefbogen in aller Welt 1.

476 Letterhead with calligraphic wordmark. Design by Oldrich Hlavsa. From J.W. Zanders, Der Briefbogen in aller Welt 1.

477 Calligraphic letterhead. Design by George Salter. From J.W. Zanders, Der Briefbogen in aller Welt 1.

478 Letterhead with handlettering. Design by Herb Lubalin. From J.W. Zanders, Der Briefbogen in aller Welt 2.

479 Wordmark by Marcel Jacno for Théâtre National Populaire. From Yusaka Kamekura, Firmen-und Warenzeichen – international. Copyright © 1965 by Zokeisha Publications Ltd, Tokyo.

480 Logo by Herb Lubalin. From Yusaka Kamekura, Firmen- und Warenzeichen – international. Copyright © 1965 by Zokeisha Publications Ltd, Tokyo.

481 Logo by Herbert Prüget for Sachsenring, Zwickau.

482 Logo by Irmgard Horlbeck-Kappler for the publishing house Philipp Reclam jun., Leipzig.

483 Logo by Herbert Prüget for Vereinigung Volkseigener Warenhäuser CENTRUM.

484 Wordmark by E. Vogenauer.

485 Logo by Gerstner & Gredinger & Kutte for Christian Holzapfel KG Buildt, furniture makers. From Yusaka Kamekura, Firmen- und Warenzeichen – international. Copyright © 1965 by Zokeisha Publications Ltd, Tokyo.

486 Logo by Kenneth R. Hollick for Amasco, Amalgamated Asphalt Co.

487 Logo by Christa Krey for the publishing house Edition Leipzig, 1963.

488 Logo by Malcolm Grear, Research & Design, Inc.

489 Logo by Eberhard Kühn for Schwarzenberg washing machines.

490 Logo by Herman Prüget for VEB Transformatorenwerk Falkensee.

491 Logo by Georg Trump for the C.E. Weber type foundry, Stuttgart.

492 Logo by Hermann Eidenbenz for Papyrus AG Kioske. From Yusaka Kamekura, Firmen-und Warenzeichen – international. Copyright © 1965 by Zokeisha Publications Ltd, Tokyo.

Figure 479

Figure 480

Figure 481

Figure 482

Figure 483

Figure 484

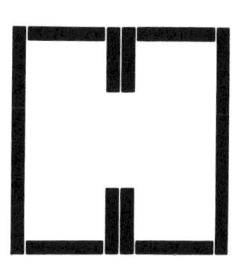

Figure 485

Figure 486

Figure 487

Figure 488

Figure 489

Figure 490

Figure 491

Figure 492

223

493 Poster for an exhibition, by Oldrich Menhart.

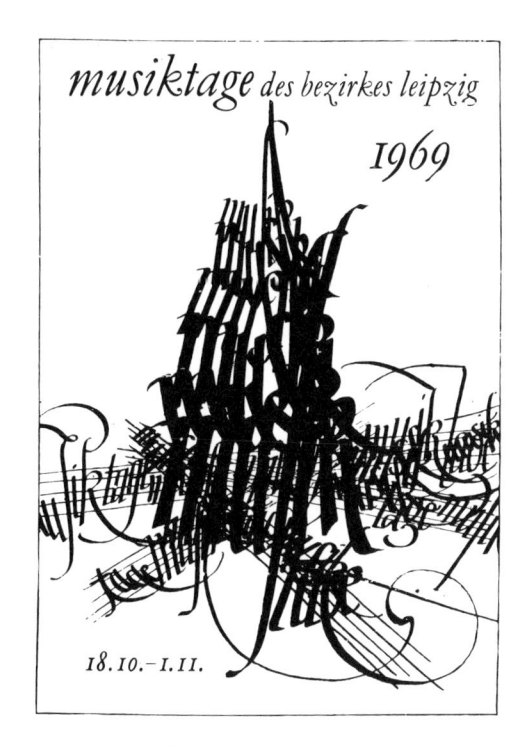

494 Poster for the Leipzig music festival, by Irmgard Horlbeck-Kappler.

495 Poster for a performance of the opera Persephone by Igor Stravinsky, by Roman Cieslewicz.

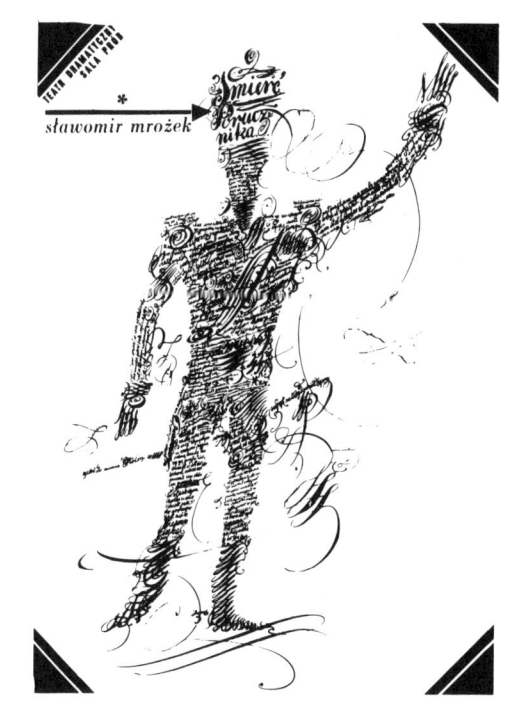

496 Poster for a performance of the Dramatisches Theaters Warschau, by F. Starowieyski.

497 Poster by Pablo Picasso.

498 Poster for the occasion of Lenin's hundredth birthday by Marius Chwedezuk.

499 Poster for a meeting of the Association Typographique Internationale.

225

226

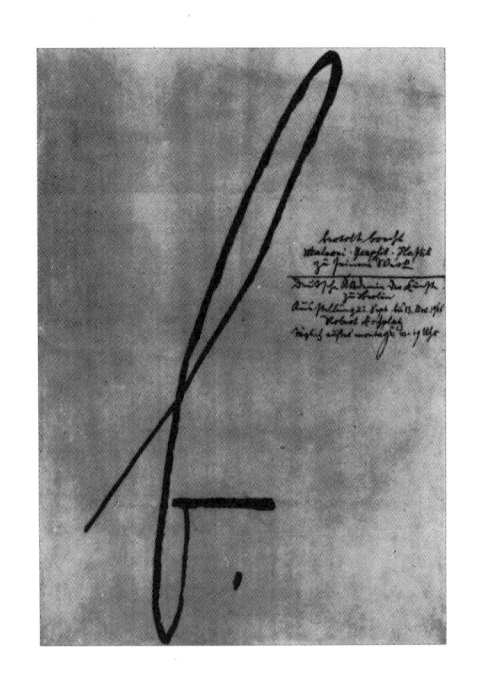

500 Poster for the third international Johann-Sebastian-Bach Competition in Leipzig, by the author.

501 Poster for an exhibition, "Bertolt Brecht's work in paintings, drawings, and sculptures," by Werner Klemke.

502 Advertisement for BIC ballpoint pens (letters in three colors) by Rudi Külling. From the catalog of the second biennial poster exhibition, Warsaw, 1968.

503 Two-color poster for the Polish circus, by Henryk Tomaszewski.

504 Multicolor poster for a furnishings exhibition, by Theo Crosby, Alan Fletcher, and Colin Forbes. From the catalog of the second biennial poster exhibition, Warsaw, 1968.

505 Two-color poster for an exhibition, by Josef Flejsar.

506 Two-color poster for an exhibition of Henry Moore's sculpture, by Henryk Tomaszewski.

507 Two-color poster for the international exhibition of book art, Leipzig, 1965, by Gert Wunderlich.

227

Figure 508

Figure 509

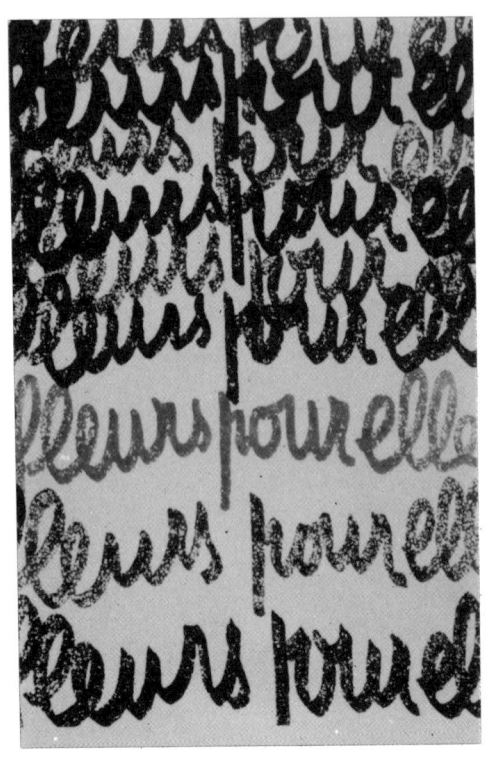

Figure 510

508 *Two-color poster for a movie, by Julian Patka.*

509 *Multicolor poster advertising "Flowers for her," by Gilbert Frankhauser.*

510 *Multicolor poster by Peter Max.*

Figure 511

Figure 512

Figure 513

Figure 514

511 Label by O.W. Hadank.

512 Label by O.W. Hadank.

513 Promotional bag for Mademoiselle magazine, by Tom Soja.

514 Package for Artone ink, by Seymour Chwast/Milton Glaser.

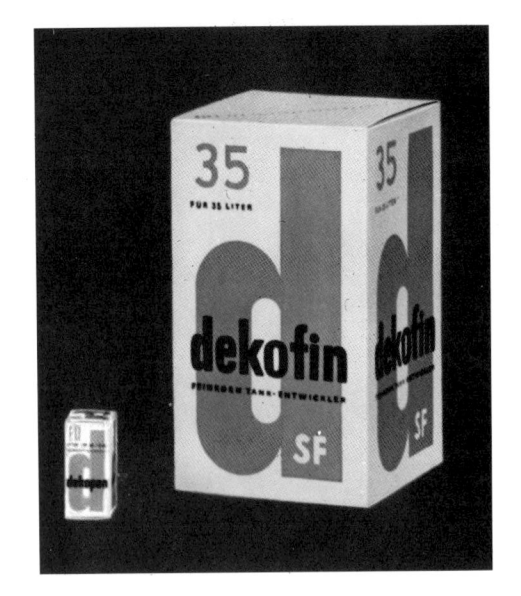

Figure 515

Figure 516

Figure 517

515 Toothpaste packaging for VEB Elbe-Chemie, Dresden.

516 Packaging for VEB Fotochemische Werke, by Klaus Wittkugel, Berlin, 1959. (Photo by Zentralinstitut für Gestaltung, Berlin.)

517 Bag made of printed wrapping paper and printed cardboard packaging for charcoal. Designed by Studio of van Winsen for J. Zorge and Son. From Crouwel/Weidemann, Packaging.

518 *Two-color jacket for* Deutsche Schriftkunst, *designed by Albert Kapr.*

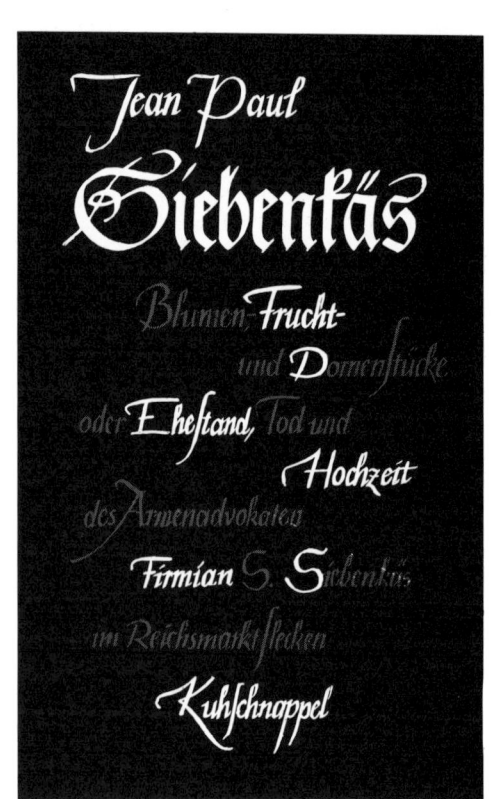

519 *Multicolor jacket by Josef Flejsar.*

520 *Multicolor jacket for Jean Paul's Siebenkäs. (Reproduction, Walter Danz.)*

521 *Multicolor jacket for a book, by Oldrich Hlavsa.*

522 *Jacket for H. Marsman's Verzamelde Gedichten, by Jan van Krimpen.*

523 *Multicolor jacket for Jonathan Swift's* Ausgewählte Werke *(Selected writings), Vol. 3,* Gulliver's Travels, *by Heinz Hellmis.*

524 *Multicolor jacket for Jean-Paul Sartre's* Die Wörter *(Words), by Sigrid Huss.*

525 *Jacket for Jaroslav Seifert's* Mozart v Praze, *by Jiri Sindler.*

526 *Jacket for* Josef Capek a Kniha, *artist unknown.*

527 *Two-color jacket for Ilya Ehrenburg's* Sturm, *by S.B. Telingater.*

528 *Jacket for HAP Grieshaber's* Totentanz von Basel, *by Albert Kapr.*

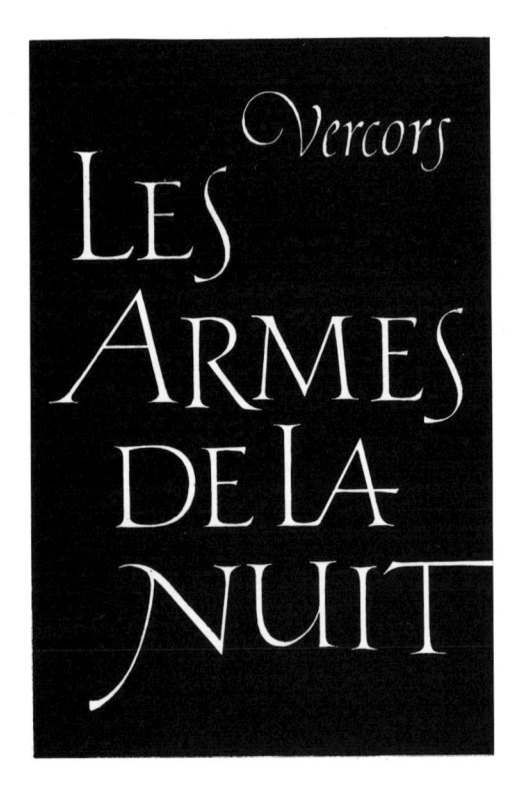

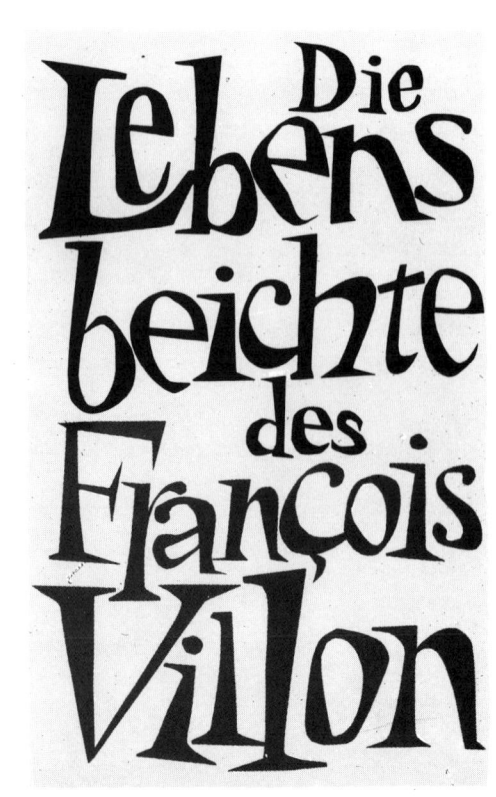

529 *Jacket for Vercors's* Les Armes de la Nuit, *by Günter Gnauck.*

530 *Two-color jacket for* Die Lebensbeichte des François Villon, *by Lother Reher.*

531 *Multicolor jacket for Jean Giraudoux's* Dramen, *by Hermann Zapf.*

532 *Multicolor jacket for Emile Zola's* Thérèse Raquin, *by Egon Pruggmayer.*

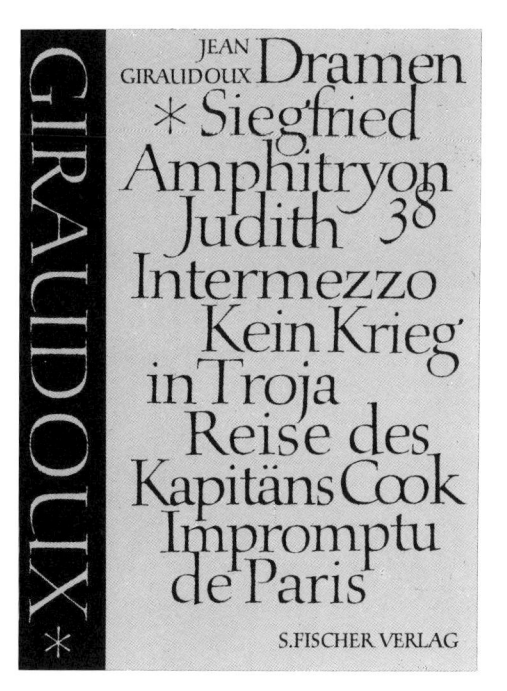

533 *Jacket for* Lästerkabinett, *by Irmgard Horlbeck-Kappler.*

534 *Multicolor jacket for Friedrich Huch's* Pitt und Fox, *by Egon Pruggmayer.*

535 *Multicolor cover for a folder,* Glück und Wohlergehen, Die Tet-Bilder aus Dong-Ho, *by Günter Gnauck.*

536 *Two-color jacket for Alexander Pushkin's* Ausgewählte Prosa *(Selected prose), by Günter Junge.*

537 *Multicolor cover for a calendar by Albert Kapr.*

538 *Cover for a book on Charles White, by Albert Kapr.*

539 *Cover for FORUM magazine, by Wolfgang Geissler.*

540 *Multicolor title page for FÜR DICH magazine, No. 28, 1970, by Thomas Schleusing.*

541 *Cover for ARCHITECTURAL REVIEW magazine.*

542 Lettering on a wall of a building in Chemnitz, by Heinz Schuman and Walter Beier. Original size 17.5 by 25.5 meters.)

543 Decorative lettering an the wall of a furniture store in Bologna.

On page 239:

544 Lettering on the foyer wall of the national theater in Mannheim, by Fritz Kühn. (Photo by Fritz Kühn.)

Figure 543

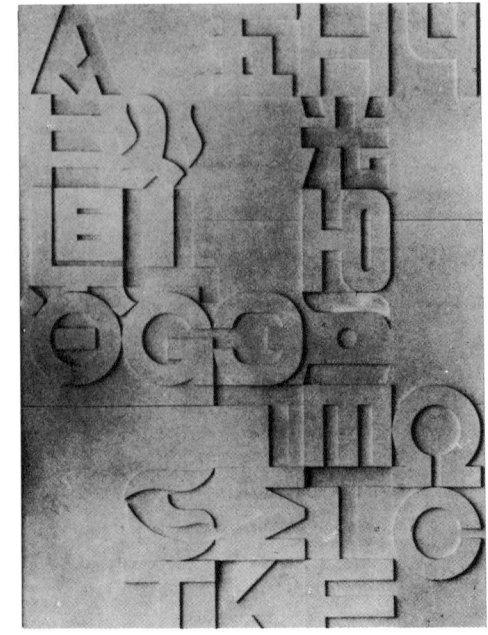

239

545 *Lettering on a wall of the offices of J.R. Geigy, Basel.*

546 Neon sign, Brühl. (Photo by Frank Schenke.)

547 Neon sign in Karl-Marx-Allee, Berlin. Design by Klaus Wittkugel. (Photo by Gunter Jasbec.)

548 *Plaque at Tower Bridge, London. From G. Prölss,* Lettering for Architects. *Stuttgart, Karl Krämer Verlag, 1957.*

549 *Lettering on the wall of The United Kingdom Pavilion, Colombo. From G. Prölss,* Lettering for Architects.

550 *Lettering for Central Junior High School, Greenwich, Connecticut, by Brownjohn, Chermayeff & Geismar. Metal letters on a brick wall.*

551 Neon sign for a cafe, Unter den Linden, Berlin. Design by Ernst Lauenroth. (Photo by Zentralinstitut für Gestaltung, Berlin.)

552 Neon sign in Leipzig, Georgiring. (Photo by Frank Schenke.)

553 Neon sign for a bookstore, Unter den Linden, Berlin. Design by Ernst Lauenroth. (Photo by Gunter Jazbec.)

554 and 555 Sign for Greyhound bus terminal, day and night. From Constantine/Jacobson, Sign Language. New York: Reinhold, n.d.)

556 Neon letters, relief set into wall, with diffused light effect. Unter den Linden, Berlin. (Photo by Gunter Jazbec.)

557 Number on a building. From Constantine/Jacobson, Sign Language.

558 Logo for the Tenth Triennale Mailand. Stencils on the street, used as signs to the exhibition area. From G. Prölss, Lettering for Architects.

Figure 551

Figure 552

Figure 553

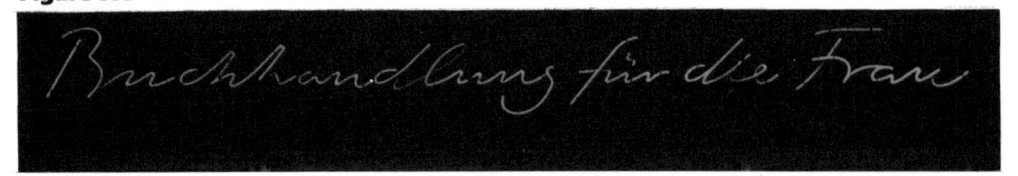

Figure 554

Figure 555

Figure 556

Figure 557

Figure 558

245

559 *Czechoslovak Pavilion, Expo 67, Montreal.*

560 *The United Kingdom Pavilion, Expo 67, Montreal.*

561 *Entrance to the Dachau Memorial, by Fritz Kühn. (Photo by Fritz Kühn.)*

Figure 562

Figure 563

Figure 564

562 and 563 Lettering on the booth of Polygrafische Industrie, Leipzig spring fair, 1969. (Photo by Doris Beinecke.)

564 Lettering on the booth of VVB Nachrichtenelektronik, Leipzig spring fair, 1969. (Photo by Doris Beinecke.)

BIBLIOGRAPHY

References

Autorenkollektiv. *Handbuch der Werbung*. Berlin: Verlag Die Wirtschaft, 1968.

Barthel, G., and C.A. Krebs. *Das Druckwerk*. Stuttgart: Verlag Berliner Union, 1963.

Behre, G. *Malerei, Schrift, Graphik in der Prazis der Webegestaltung*. Ulm-Söflingen: Karl Gröner Verlag, 1954.

Braune, H. *Das Skizzieren von Satzschriften*. In "Papier und Druck," 8/66. Leipzig: VEB Fachbuchverlag, 1966.

Buchartz, M. *Gleichnis der Harmonie*. Munich: Prestel-Verlag, 1955.

Doede, W. *Schön schreiben, eine Kunst*. Munich: Prestel-Verlag, 1957.

Der Große Duden: Leitfaden der deutschen Rechtschreibung und Zeichensetzung mit Hinweisen auf grammatische Schwierigkeiten Vorschriften für den Schriftsatz. Leipzig: VEB Bibliographisches Institut, 1967.

Gollwitzer, G. *Kleine Zeichenschule*. Berlin: Volk und Wissen Volkseigener Verlag, 1958.

Gropius, W. *Architektur*, second edition. Frankfurt/ M. — Hamburg: Fischer-Bücherei, 1959.

Johnston, E. *Writing Illuminating and Lettering*. New York: Taplinger, 1977.

Kaech, W. *Rhythmus und Proportionen in der Schrift*. Olten und Freiburg im Breisgau: Walter-Verlag, 1956.

Kallich, H. *Werbung und Druck. Ein Ratgeber für die Herstellung von Drucksachen*. Berlin: Verlag Die Wirtschaft, 1962.

Kandinsky, W.W. *Point and Line to Plane*. New York: Dover, 1979.

Kapr, A. *The Art of Lettering*, Munich, New York, London, Paris, 1983.

Kapr, A. *Deutsche Schriftkunst*, second edition. Dresden: VEB Verlag der Kunst, 1959.

Kapr, A. *Fundament zum rechten Schreiben. Eine Schriftfibel*. Leipzig: VEB Fachbuchverlag, 1958.

Kapr. A. *Buchgestaltung*. Dresden: VEB Verlag der Kunst, 1963.

Kapr, A. *Die Klassifikation der Druckschriften*. In "Schriftmusterkartei. Ratschläge zur praktischen Anwendung der Schriftmuster- karteikarten." Leipzig: VEB Fachbuchverlag 1967.

Kapr, A. *Probleme der typografischen Kommunikation*. In "Beiträge zur Grafik und Buchgestaltung." Sonderdruck der Hochschule für Grafik und Buchkunst. Leipzig: Hochshule für Grafik und Buchkunst, 1964.

Lebek, J. *Holzschnittfibel*. Dresden: VEB Verlag der Kunst, 1962

Luers, H. *Das Fachwissen des Buchbinders*, fifth edition, Stuttgart: Buchbinder-Verlag Max Hettler, n.d.

Moessner, G. *Was Setzer, Drucker und Verlagshersteller von der Buchbinderarbeit wissen sollten*. Stuttgart: Buchbinder-Verlag Max Hettler, 1960.

Muller, W., and A. Enskat. *Theorie und Praxis der Graphologie I*. Rudolstadt: Greifenverlag, 1949.

Muzika, F. *Die Schöne Schrift*, Vols. 1 and 2. Prague: Artia Verlag, 1965.

Nettelhorst, L. *Schrift muss passen*. Fachbuchreihe Wissenschaft und Werbung Essen: Wirtschaft und Werbung, Verlagsgesellschaft mbH, 1959.

Prölss, G. *Schrift fur Architekten*. Stuttgart: Verlag Karl Krämer, 1957.

Renner, P. *Ordnung und Harmonie der Farben*. Ravensburg: Otto Maier Verlag, 1947.

Rossing, R. *Schrift und Foto*. In "neue werbung," 1/67. Berlin: Die Wirtschaft, 1967.

Rossner, C. *Schriftmischen*. In "Schriftmusterkartei. Ratschläge zur praktischen Anwendung der Schriftmuster-Karteikarten." Leipzig: VEB Fachbuchverlag, 1967.

Ruder, E. *Die richtige Schriftwahl*. In "Hausmitteilungen der Linotype GmbH Berlin und Frankfurt am Main," Heft 56, November 1962.

Ruder, E. *Typography: A Manual of Design*. Teufen AR: Verlag Arthur Niggli, 1967.

Schenk, W. *Die Schriften des Malers*. Giessen: Fachbuchverlag Dr. Pfannenberg, 1958.

Schenk, W. *Die Schrift im Malerhandwerk*. Berlin: Verlag fur Bauwesen, 1965.

Schneidler, E. *Der Wassermann. Ein Jahrbuch fur Buchermacher. Uber Forschungen im Bereiche des Schreibens und des Schriftentwurfs, des Setzens, der Bildgestaltung, der Bildwiedergabe und des Druckens*. Sonderdruck der Akademie der bildenden Kunste. Stuttgart: Akademie der bildenden Kunste, 1948.

Schriftmusterkartei, second edition. Leipzig: VEB Fachbuchverlag, 1968.

Tschichold, J. *Treasury of Alphabets and Lettering*. Reprint. New York: Design Press, 1992, and London: Lund Humphries, 1992.

Tschichold, J. *Willkurfreie Massverhaltnisse der Buchseite und des Satzspiegels*. In "Druckspiegel," Typographische Beilage 7a/1964.

Zentrale Werbeabteilung der Leder- und Druckfarben GmbH, Leipzig: Rationelles Farbmischen, Buntfarben fur den Buchdruck, Buntfarben fur den Offsetdruck.

Selected Further Reading

Aldis, Harry G. *The Printed Book*. Cambridge, England, 1951.

Bartram, Alan. *The English Lettering Tradition*. London, 1986.

Bigelow, Charles, et al. (ed.). *Fine Print on Type: The Best of Fine Print Magazine on Type and Typography*. San Francisco/London, 1988.

Carter, Harry. *A View of Early Typography up to about 1600*. Oxford, 1968.

Child, Heather (ed.). *Formal Penmanship and Other Papers: Edward Johnston*. London, 1971.

Child, Heather. *The Calligraphers Handbook*. New York, 1986.

Child, Heather, and Justin Howes. *Lessons in Formal Writing: Edward Johnston*. London, 1986.

Craig, James. *Production for the Graphic Designer*. New York, 1974.

Day, Kenneth (ed.). *Book Typography 1815–1965 in Europe and the United States of America*. London, 1966.

DeVinne, Theodore Low. *Modern Methods of Book Composition: A Treatise on Type-setting by Hand and by Machine and on the Proper Arrangement and Imposition of Pages*. New York, 1904.

Dreyfus, John (ed.). *Type Specimen Facsimiles: Reproductions of Fifteen Type-specimen Sheets issued between the Sixteenth and Eighteenth Centuries*. London, 1963.

Dreyfus, John (ed.). *Type Specimen Facsimiles II: Reproductions of Fifteen Type-specimen Sheets issued between the Sixteenth and Eighteenth Centuries*. London, 1972.

Folsom, Rose. *The Calligrapher's Dictionary*. London, 1990.

Gill, Eric. *An Essay on Typography* (rev. ed.). London/Boston, 1988.

Gray, Bill. *Calligraphy Tips*. New York, 1989.

Gray, Bill. *Lettering Tips*. New York, 1980.

Gray, Bill. *Studio Tips*. New York, 1976.

Gray, Bill. *Tips on Type*. New York, 1983.

Johnson, A.F. *Type Designs: Their History and Development*. London, 1966.

Jaspert, W. Pincus, W. Turner Berry, and A.F. Johnson. *The Encyclopaedia of Type Faces* (rev. ed.). London, 1983.

Johnston, Edward. *Writing and Illuminating and Lettering* (rev. ed.), ed. by W.R. Lethaby. London, 1944; New York, 1977.

McLean, Ruari. *Modern Book Design from William Morris to the Present Day*. London, 1958.

McLean, Ruari. *Jan Tschichold: Typographer*. London/Boston, 1975.

McLean, Ruari. *The Thames and Hudson Manual of Typography*. London, 1980.

Modern Scribes & Lettering Artists. London, 1990.

Morison, Stanley. *On Type Designs Past and Present: A Brief Introduction*. London, 1962.

Morison, Stanley. *First Principles of Typography* (2nd ed.). Cambridge, England, 1967.

Rookledge, Gordon, and Christopher Perfect. *Rookledge's International Typefinder*. London, 1983.

Simon, Oliver. *Introduction to Typography*. London, 1969.

Spencer, Herbert. *Pioneers of Modern Typography* (rev. ed.). London/New York, 1982.

Stone, Bernard, and Arthur Eckstein. *Preparing Art for Printing*. New York, 1983.

Sutton, James, and Alan Bartram. *An Atlas of Typeforms*. London, 1968.

Tschichold, Jan. *Asymmetric Typography*. London/Toronto/New York, 1967.

Tschichold, Jan. *Treasury of Alphabets and Lettering* (reprint). New York/London, 1992.

Tschichold, Jan. *The Form of the Book: Essays on the Morality of Good Design*. London/Washington, 1992.

Twyman, Michael. *Printing 1770–1970: An Illustrated History of its Development and Uses in England*. London, 1970.

Updike, Daniel Berkeley. *Printing Types, their History, Forms and Use: A Study in Survivals* (rev. ed.). London/New York, 1980.

Wills, F.H. *Fundamentals of Layout*. New York, 1965.

Wilson, Adrian. *The Design of Books*. New York, 1974.

Wallis, L.W. *Modern Encyclopedia of Typefaces 1960–90*. London, 1990.

Zeier, Franz. *Books Boxes and Portfolios*. New York, 1990.

INDEX

Akzidenz-grotesk *152, 153, 174*
Amati neoclassical roman *120*
Arabic numerals *106*
Architecture, lettering and *72, 171, 205–212*
Artists' works, handlettered book *179*
Athenaeum, neoclassical roman *120*

Ball-point pens *64, 197*
Banners, lettering *211*
Barbedor, Louis *125*
Baroque fraktur *101*
Baroque italic *209*
Baroque roman *119, 120*
Baroque script *125–129*
Baskerville, John *119*
Baskerville roman *119*
Batarde styles *70, 95–97*
Bayer, Herbert *42*
Bellyband, book jacket *194*
Bembo italic *118*
Bembo roman *106, 109*
Birthday cards, lettering *171*
Bodoni, Giambattista *13, 120*
Bodoni roman *120, 122, 172, 195, 206, 212*
Bohemian batarde *95*
Bold roman neoclassical type *130, 131–137*
Book, handlettered
    generally *179–187*
    binding of *187–190*
    colophon *186*
    format *180*
    front matter *185–186*
    layout *186–187*
    page layout *180–183*
    text, design *183*
    typeset book compared *180*
Book covers, handlettered book *188, 190*
Book jackets, design *171, 194*
Bookbinding
    generally *187–190*
    adhesive binding *190*
    covers *188, 190*
    folders *188*
    multiple pages *188–190*
    paper, grain of *187*
    paperfolding *188*
    parchment for *176*
    side-sewing *189*
    simple binding *188*
Brecht, Bertholdt *28*
Breitkopf fraktur *101*
Brochures, format *29*
Bronze plaques, lettering on *211*
Brushes
    drawing *37*
    lettering *37, 41, 175*
Brushwork, lettering *36, 41, 42, 50, 51*

Calligraphy
    generally *174–189*
    book, see Book, handlettered
    engraving *125*
Cancellaresca *112*
Capital letters
    decorative *13, 98, 102–103, 126–127, 174*
    initial capitals, *see* Initial capitals
    italic script *62*
    monumental roman *121, 123*
    Renaissance capital, construction *42*
    roman *13, 70, 72, 80, 90, 106*
    rustic capitals *82–83, 88, 90*
    sans serif roman *42–51*
Carolingian miniscule *20, 88, 95, 106*
Caslon, William *119*
Caslon, William IV *130*
Caslon type *110*
Centaur type *106*
Certificates
    calligraphy *175*
    tubes for *188*
Chalk *37, 175*
Chancery cursive script *105, 112*
Chinese binding, handlettered book *189*
Chinese ink *33, 176, 197*
Chinese roman type *109*
Circular arrangement, letters of logo *192*
Civilite typé *95, 99*
Clarendon type *130*
Coats of arms, lettering *175*
Color
    associations *17*
    calligraphy, documents and short texts *175*
    layout *28*
    light/dark contrast *16*
    pure and "broken" color *17*
    warm/cool contrast *17*
    woodcut printing *203*
Column structure, handlettered book *183*
Company history, handlettered book *179*
Composition, design *26–28*
Computer typesetting *196, 199, 200*
Condensed sans serif roman
    see also Sans serif roman
    display type *195*
    drawing and cutting letters *39–41*
    examples *160–165*
    lettering with flat brush *41, 42, 51*
    mixing with other type *174*

Contrast, design element *13, 14, 16, 17, 19, 26, 28, 172*
Corvinius, neoclassical roman 120
Cursive script, *see* Italic script
Cursive typeface, *see* Italic type
Cyrillic letters and type *165–167*

dePolanco, Juan *125*
Design, elements of *13–19*
Desmoulins, François *125*
Didot roman type *120*
Diotima, neoclassical roman *120*
Diplomas, calligraphy *175*
Display type *130 et seq.*
Documents
    calligraphy *175*
    tubes for *188*
Drawing
    brushes and other material for *37*
    pens for *33–35*
Dürer, Albrecht *42, 100, 101*

Eckmann type *150*
Egizio type *130*
Egyptian italic *146, 148*
Egyptian style type *20, 130, 139–143, 146, 195, 106*
"Elephant trunk" script *95, 98*
English binding, handlettered book *190*
English scripts *35, 105, 125, 128, 174*
Engraving
    calligraphy *125*
    method *204*
Exhibitions, lettering for *194, 195*

Faust italic *118*
Felt-tip markers *37, 51, 64*
Figgins, Vincent *130*
Futura, neoclassical roman *120*
Fleischmann, Johann Michael *119*
Focal points, movement, design *18, 19*
Folio type *130*
Folios, handlettered book *183*
Footnotes, handlettered book *183*
Footnotes, handlettered *184*
Form, design *13*
Format
    handlettered book *179, 187*
    paper, standard sizes and *29, 30*
Fournier, Pierre Simon *119*
Fraktur
    generally *98–104*
    hyphenation *30*
    line spacing *25*
    mixing with other type *174*
    stone carving *209*
Frank, Paul *13, 102, 103*
French batarde script *95*
French Renaissance roman type *106*
Futura type *151*

Garamond type *106, 109, 172, 174 195, 206*
Garamond italic *118*
German batarde script *98*
Gilding, raised *176–179*
Gill Sans *209*
Glass, lettering on *211*
Gold leaf, raised gilding *176–179*
Gold, lettering documents *175*
Gouache *33, 197*
Gothic letters *70, 90–105*
Gothic scripts and variations *95–105, 112*
Grandjean, Austin *119*
Granjon, Robert *112*
Graphic artwork
    photographs, lettering on *197*
    preparation for reproduction *196, 197*
    resist technique, lettering *204*
    scratch technique, letteing *205*
    type rough layouts *196, 197*
Gravestones, carving on *209, 210*
Griffo, Francesco *112*
Gropius, Walter *13*
Gutenberg, Johannes *90, 106*

Handlettered book, *see* Book, handlettered
Handwriting, personal, italic script and, *62, 63, 64, 84, 95, 101, 105, 112, 125*
Headings, handlettered book *183*
Helvetica type *130, 174, 195*
Hirmer, Helmut *177*
Historical associations, typefaces *173*
Hoefer, Karlgeorg *13*
Hyphenation, composition *30*

Illustrations, type and *172*
Imprimatur roman *119*
Ingres paper *33*
Initial capitals
    see also Capital letters
    book, handlettered *183*
    documents and short texts *175*
    layouts *13, 27, 28*
Ink
    Chinese ink *33, 176, 197*
    lettering and drawing ink *33*
    reservoirs, pen nibs *35*
    rubber stamp letters, ink for *195*
    woodcut printing *202*
Italic script
    calligraphy *174, 175*
    Cyrillic alphabet *166*
    even-stroke script *66*
    handwriting personal, and *62, 63 64, 84, 94, 101, 105, 112, 125*
    history *58, 70, 84–89, 112, 114, 125*
    lowercase letters, practice exercise *57–61*
    marginalia *184*
    pens for *64*
    uppercase letters *62*
Italic type *112*
Italienne type styles *130, 145*

Jannon roman *106*
Japanese binding, handlettered book *189*
Jenson roman *106*
Johnston, Edward *176, 177, 178*
Justified text *182*

Kandinsky, W.W. *19*
Kapr, Albert *17, 71, 100, 130, 182, 183, 184*
Kis, Nicholas *119*
Koch, Rudolf *20, 200*
Kurrent type *105*

Labels
    design *194*
    format *29, 30*
Large-scale letters *196, 211, 212*
Layout
    asymmetrical *26*
    color *28*
    handlettered book *186–187*
    initial capitals *27, 28*
    page layout, handlettered book *180–183*
    symmetrical *26*
Leipzig roman *110*
Lettering
    generally *32*
    exercises *39–68*
    materials for *33*
    spacing, composition *23–24*
    styles, history of *70, 71*
    tools for *34*
    work surface *33*
Line spacing, composition *25–26, 182*
Liner brushes, lettering *51*
Linoleum, letters cut from *51, 203–204*
Logotype *190–192*
Lombard versals *90, 92*
Lowercase letters
    even-stroke sans serif *52*
    history of *70, 84–89*
    italic script *58–61*

Lucas, Francisco *112*

Mahlstick, use of *37, 41, 51*
Manutius, Aldus *106, 112*
Marginalia *184*
Margins, handlettered book *182*
Monumental capitals, roman *13, 72–81, 111, 207, 209–210*
Movement, in visual compositions *18*
Murals, lettering, sgraffito *206, 208*
Muzika, Frantisek *71, 88, 95, 106, 130*

Neoclassical roman *120, 174*
Neon signs, lettering *207–208*
Neudörffer, Johann the Elder *13, 98, 100, 101*
Neutra type *130, 138*
Numerals
    Arabic *106*
    italic script *62*
    roman sans serif *57*

Offset printing ink, rubber stamp letters, *195*
Optical illusions
    color design *16*
    design *14, 16*
    lettering *20–23*

Pacioli *42*
Packaging, design *171, 193, 194*
Page numbers, handlettered book *183*
Palatino, Giambattista *58, 96, 106, 112*
Paper
    see also Parchment
    grain of, bookbinding *187*
    lettering and drawing, paper for *33*
    letters cut from *51*
    sizes, standard; format and *29, 30*
Paperfolding, handlettered book *188*
Pragraphs, handlettered book *183*
Parchment
    see also Paper
    bookbinding, preparation for *176*
    calligraphy, choice for *175, 176*
Pencil, drawing *37*
Pens and nibs
    lettering, choice for *34*
    pointed, calligraphy *125*
    reed pens *34, 175*
    round-ended nibs, lettering with *48*
    steel pens *35, 175*
Pergamon, neoclassical roman *120*
Photographs
    lettering on, artwork for reproduction *197*
    type and *172*
Phototypesetting *198, 199*
Poetry, handlettered book *179, 180, 184*
Postage stamps *171*
Posters, design *171, 192–193*
Printing typefaces, history *106*
Psychological associations, typefaces *173*
Punctuation marks
    composition *30*
    handlettered books *184*

Quadrata *84, 106*
Quill pens *34, 88, 125*

Raised gilding *176–179*
Readability, design for *13, 16*
Renaissance italic *58, 64, 106, 112, 114*
Renaissance roman *70, 106, 119, 120, 174*
Renner, Paul *173*
Resist technique, lettering *204*
Rhythm of movement, design *13–14*
Rockner, Vincenz *101*
Roman letters
    Baroque roman *119, 120*
    capitals *13, 70, 72, 80, 90 106*
    cursives *84*
    neoclassical roman *120, 174*
    quadrata *84*
    sans serif, *see* Condensed sans serif roman; Sans serif roman
Rotunda letterforms *70, 90, 93, 94*
Round batarde script *95*
Rubber stamps, lettering *195–196*
Ruder, Emil *13, 18*
Rustic capitals *82–83, 88, 90*

Sans serif roman
    architecture, use in *206*
    constructing or drawing letters *42*
    display type *130, 149–164*
    even-stroke widths *42–51, 52*
    large-scale letters, banners *212*
    lowercase letters, varying stroke widths *57*
    mixing with other type *174*
    uppercase letters *42–51*
    varying stroke widths, drawing exercise *53–57*
Schneidler, Ernst *13, 19, 101, 200*
Schönsperger *101*
Schwabacher script *98, 174*
Scratch technique, lettering *205*
Script brushes, lettering *51*

Scrolls, handlettered documents *188*
Semper, Gottfried *208*
Sgraffito *206, 208*
Size relationships, design *14–15*
Spacing *see* Layout
Spanish italic type *112, 114*
Speeches, handlettered book *179*
Spiked batarde script *96*
Square capitals, roman *84*
Standard, Paul *65*
Stenciling, letters *196*
Stentor type *117*
Stone carving, letters *13, 72–81, 111, 207, 209–210*
Store design, lettering *205*
Subheadings, handlettered book *183*

Tagliente, Giovanni Antonio *96, 112*
Tempera, drawing with *33*
Text
    design, handlettered book *183–185*
    layout, handlettered book *182*
    short texts, calligraphy *175*
Textural letters *25, 70, 90, 95, 98, 174*
Thorne, Robert *130*
Three-dimensional effects, composition *19*
Tiemann, neoclassical roman *120*
Times type *119*
Title page, handlettered book *185–186*
Tost, Renate *65*
Tracing paper *37*
Trajan's column, letterforms *13, 72, 209*
Transfer type *196*
Trump mediaeval type *106*
Tschichold, Jan *13, 30, 32, 114, 180, 182*
Tschörtner type *106, 174*
Tubes, for handlettered documents *188*
Tuscan type *130, 144*
Typography
    choice of typefaces *172, 173*
    computer typesetting *196, 199, 200*
    mixing typefaces *173–174*
    phototypesetting *198, 199*

Uncials and half uncials *70, 86, 88, 90, 166*
Univers type *95, 130*
Upper Rhine batarde *98*
Uppercase letters, *see* Capital letters

van Dyck, Christoph *119*
van den Velde, Jan *125*
Vellum, *see* Parchment
Venetian typefaces *106*
Vincentino, Ludovico (Arrighi) *110, 112*
Volta type *130*

Wagner, Leonhard   *98, 101*
Walbaum roman   *120, 122, 172, 206*
Wallau   *90*
Walls, lettering on   *206, 208*
Watercolor brush, lettering   *51*
Weiss roman   *106*
Wittenberg fraktur   *98*
Wood, writing tool   *37, 51*
Woodcut letters and lettering   *51, 98, 200–203*
Wood spacing   *25*
Work surface, requirements   *33*

Zapf, Hermann   *104*
Zentenar fraktur   *101*